Oil and Gas Corrosion Prevention

Oil and Gas Corrosion Prevention

From Surface Facilities to Refineries

James G. Speight PhD, DSc
CD&W Inc.,
Laramie, Wyoming, USA

AMSTERDAM • BOSTON • HEIDELBERG • LONDON
NEW YORK • OXFORD • PARIS • SAN DIEGO
SAN FRANCISCO • SINGAPORE • SYDNEY • TOKYO

Gulf Professional Publishing is an imprint of Elsevier

Gulf Professional Publishing is an imprint of Elsevier
25 Wyman Street, Waltham, MA 02451, USA
The Boulevard, Langford Lane, Kidlington, Oxford, OX5 1GB, UK

Notices
Knowledge and best practice in this field are constantly changing. As new research and experience broaden our understanding, changes in research methods or professional practices, may become necessary.

Practitioners and researchers must always rely on their own experience and knowledge in evaluating and using any information or methods described herein. In using such information or methods they should be mindful of their own safety and the safety of others, including parties for whom they have a professional responsibility.

To the fullest extent of the law, neither the Publisher nor the authors, contributors, or editors, assume any liability for any injury and/or damage to persons or property as a matter of products liability, negligence or otherwise, or from any use or operation of any methods, products, instructions, or ideas contained in the material herein.

Library of Congress Cataloging-in-Publication Data
A catalog record for this book is available from the Library of Congress

British Library Cataloguing-in-Publication Data
A catalogue record for this book is available from the British Library.

ISBN: 978-0-12-800346-6

For information on all Gulf Professional Publishing publications
visit our website at store.elsevier.com

This book has been manufactured using Print On Demand technology. Each copy is produced to order and is limited to black ink. The online version of this book will show color figures where appropriate.

ELSEVIER   Book Aid International   **Working together to grow libraries in developing countries**

www.elsevier.com • www.bookaid.org

CONTENTS

PREFACE

Corrosion can occur at any point or at any time during petroleum and natural gas recovery and processing. Thus is particularly true in ageing refineries and gas processing plants which can present a serious safety hazard.

Key findings into corrosion processes in refinery equipment and gas processing plants as well as corrosion in offshore structures are presented.

This book summarizes the key corrosion processes in refinery and gas processing equipment – such as storage tanks, reactors, and sour water strippers – and how it can be measured and controlled. Methods of testing for corrosion are presented as well as preventative measures. The book also contains a helpful glossary that will assist the reader in understanding the terminology of corrosion.

Dr. James G. Speight
Laramie, Wyoming
December 2013

CHAPTER 1

Corrosion

Note: This chapter is available on the companion website: http://store. elsevier.com/product.jsp?isbn=9780128003466&_requestid=1050998.

1.1 ABSTRACT

Corrosion is the deterioration of a material as a result of its interaction with its surroundings and can occur at any point or at any time during petroleum and natural gas processing. Although this definition is applicable to any type of material, it is typically reserved for metallic alloys. Furthermore, corrosion processes not only influence the chemical properties of a metal or metal alloys, but they also generate changes in their physical properties and mechanical behaviors.

It is the purpose of this chapter to present the terminology of the different types of corrosion, as well as the key chemical aspects of corrosion processes and the types of corrosion that can occur in a refinery or gas processing plant.

1.2 CONTENT

CHAPTER 2

Materials of Construction for Refinery Units

2.1 INTRODUCTION

To accomplish the conversion of crude oil feedstock to saleable products, a refinery is a large industrial complex with extensive piping to transport different fluids between large reactor units (Speight and Ozum, 2002; Hsu and Robinson, 2006; Gary et al., 2007; Speight, 2014). The complexity of refineries depends on the type of crude oil being processed and the type of products being manufactured. Oil refineries typically process between 100,000 and 2,000,000 barrels of crude oil per day into saleable petroleum products. Because of this high-throughput capacity, most refinery processing units are operated continuously, which makes process optimization and process control very desirable.

The relative complexity of a refinery is determined by the number and nature of the process units. Refining begins with the desalting/dewatering of the petroleum feedstock(s) followed by fractionation (distillation) into separate boiling-range fractions: the resultant products are directly related to the properties of the crude oil. Beyond the desalting operation, refineries are classified into (1) simple refineries, which include distillation, catalytic cracking, and distillate hydrotreating units, (2) complex refineries, which—in addition to the units at simple refineries—include catalytic reforming, alkylation or polymerization units, and gas processing operations, and (3) very complex refineries, which—in addition to the units found in complex refineries—include a hydrocracking unit, a coking unit, an asphalt plant, and—often—a petrochemical plant (Speight and Ozum, 2002; Hsu and Robinson, 2006; Gary et al., 2007; Speight, 2014). Each unit in the refinery is subject to common corrosion problems (Table 2.1). From the refinery, the products are transported to product distribution centers or terminals. Pipelines are used to transport petroleum to the refinery and also to transport petroleum products to the various distribution centers.

Metallic materials used to manufacture equipment for the refining and gas processing industries are subjected to a wide variety of potential damage mechanisms, the most common being corrosion (Chapter 1).

Table 2.1 Typical Refinery Units, Materials of Construction, and Operating Conditions

Unit	Material	Temperature (°C)	Pressure** (psi)	Corrosion Rate** (mpy)	Typical Corrosion Type	Effect Primarily due to
Desalter	Carbon steel	50	50	200	Localized pitting corrosion	Salt
Atmospheric distillation	Carbon steel, Cr–Mo steels, 12Cr, 316 stainless steel, Monel, and 70 – 30 copper/nickel alloy	371	50	315	Localized pitting corrosion, and flow-induced localized corrosion	Naphthenic acid and sulfur, HCl in overhead
Vacuum distillation	Carbon steel, 9Cr–1Mo steel, and austenitic stainless steel	400	10	~417	Localized pitting corrosion	Naphthenic acid, sulfur, HCl in overhead
Catalytic cracking	Carbon steel and stainless steel with refractory lining, Inconel 625, alloy 800	600	100		Intergranular SCC, graphitization, erosion	
Hydrotreating	Carbon steel, Cr–Mo steels, alloy 825, 321 stainless steel, 347 stainless steel, alloy 800, alloy 800H	670	2000	~137	SSC, SCC, hydrogen flaking, pitting corrosion	H_2S, polythionic acid, and ammonium salts
Hydrodesulfurization	Carbon steel, 316L stainless steel, 405 stainless steel, alloy 825, 9Cr–Al, and graphitized SA 268	593	750	383	Intergranular cracking, localized pitting corrosion	H_2S
Catalytic reforming	Carbon steel and 2.25Cr 1Mo steel	650	360	48	Metal dusting, carburization, and localized pitting corrosion	Chloride, ammonia, caustic
Visbreaker	Carbon steel	220		16		
Coker	Carbon steel	300		20	High-temperature oxidation and sulfidation	H_2S
Alkylation	Carbon steel, alloy 400, and Monel 400	188	60	100	Localized pitting corrosion	SO_2 and acid (sulfuric and hydrofluoric acid)
Gas treating	Carbon steel	128	1250	10	Localized pitting corrosion	H_2S, CO_2, amine
Sour water stripper	Carbon steel, 316L stainless steel, alloy 825, Ni–alloy C-276, alloy 2205, alloy 2507 and grade 2 titanium	245	100	85	Localized pitting corrosion, erosion-corrosion	H_2S, flow, and chloride
Sulfur recovery	Carbon steel, 304L stainless steel, refractory	121		16	Localized pitting corrosion	H_2S

SCC, stress corrosion cracking.

Safe operation of a refinery or gas processing plant depends on understanding these corrosion degradation mechanisms, making the proper material selection, devising corrosion control, creating inspection programs for early detection of problems, and monitoring material performance.

2.2 METALS AND ALLOYS

There are many diverse applications for the stainless and heat-resistant alloys throughout the range of temperatures encountered in the refining, petrochemical, and gas processing industries (Tillack and Guthrie, 1998). A wide variety of iron- and nickel-based materials are used for pressure vessels, piping, fittings, valves, and other equipment in refineries and petrochemical plants. The most common of these is carbon steel in which the main interstitial constituent is carbon in the range of 0.12−2.0% w/w. The term *carbon steel* may also be used in reference to steel that is not stainless steel. As the percentage of carbon increases, steel has the ability to become harder and stronger through heat treating, but this is accompanied by a lowering of the ductility along with a lower melting point and lower weldability.

Furthermore, drilling for crude oil and natural gas, as well as the production, refining, storage, and transportation inflicts corrosion on the equipment. For example, acid-bearing fluids used during drilling operations can corrode the tubing through which they flow. The presence of hydrogen sulfide, other sulfur compounds, and mineral matter can induce corrosion in pipelines either chemically or through attrition— problems that have affected many pipelines transporting crude oil and natural gas from the wellhead to refineries.

2.2.1 Metals

Metals used for unit constriction should be durable and corrosion resistant. Most refining and petrochemical processing equipment—as well as gas processing equipment—is designed and fabricated to the requirements of the American Society of Mechanical Engineers (ASME) or equivalent pressure vessel and piping codes of other countries. These codes establish the basis for and the setting of allowable stresses. Thus, the mechanical properties of a material are usually the first criteria used in the selection process. This is especially important for applications at temperatures in the creep range in which a minor difference in operating temperature can significantly affect the load-carrying ability of the material.

Without adequate corrosion resistance (or corrosion allowance), the component will fall short of the minimum design life desired. In the refining and petrochemical industries, this is typically set at 10 years or more. The additional cost usually associated with choosing increased corrosion resistance during the selection process is invariably less than that due to product contamination or lost production and high maintenance costs due to premature failure.

However, unlike mechanical properties, there are very few, if any, codes governing corrosion properties of metals. For some applications or services, recommended practices have been published by organizations such as the American Petroleum Institute and NACE International. Small variations in the composition of a process stream or in operating conditions can cause very different corrosion rates. Therefore, the most reliable basis for material selection is operating experience from similar plants and environments or from pilot plant evaluations.

2.2.2 Alloys

Generally, wrought and cast heat-resistant alloys are required in numerous refinery and petrochemical applications because of the combination of aggressive environments and strength requirements. High levels of nickel and chromium provide alloys with the capability of fulfilling these requirements on an economical basis. The alloys used for reformer tubing have been going through a transformation from HK-40 (an austenitic Fe−Cr−Ni alloy that has been a standard heat-resistant material for several decades) to steel-Nb for tubes and, more recently, to the micro-alloyed grades. This evolution is continuing as reflected by the increasing use of 35Cr−45Ni (35% w/w chromium−45% w/w nickel) alloys.

2.2.2.1 Ferrous Alloys

Ferrous alloys are alloys that contain iron as a principal element and are conveniently subdivided into three categories: (1) cast iron, (2) ferroalloys, and (3) steel. On the other hand, a nonferrous alloy is an alloy that does not contain iron.

Cast iron is an iron or a ferrous alloy that has been heated until it liquefies and is then poured into a mold to solidify. The alloy constituents affect the color when fractured: white cast iron has carbide impurities that allow cracks to pass straight through, while gray cast iron has graphite flakes that deflect a passing crack and initiate countless new cracks as the material breaks. Carbon (C) and silicon (Si) are the main alloying elements, with the amount ranging from 2.1% to 4% w/w and 1% to 3% w/w, respectively.

Ferroalloy refers to various alloys of iron with a high proportion of one or more other elements such as manganese, aluminum, or silicon (Fichte, 2005). It is used in the production of steel and alloys and imparts distinctive qualities to steel and cast iron or serves important functions during production. It is, therefore, closely associated with the iron and steel industry, the leading consumer of ferroalloys.

Steel is the common name for a large family of iron alloys and can either be cast directly to shape or into ingots that are reheated and hot worked into a wrought shape by forging, extrusion, rolling, or other processes. It the most important alloy used in every part of the oil and gas industry: from production and processing to the distribution of refined products.

Wrought steel is the most common engineering material used, and it comes in a variety of forms with different finishes and properties. Stainless steels are high-alloy steels that have superior corrosion resistance to other steels because they contain large amounts of chromium. Stainless steel contains at least 10% w/w chromium, with or without other elements. Stainless steel can be divided into three basic groups based on their crystalline structure: (1) austenitic, (2) ferritic, and (3) martensitic. Mild steel is an alloy containing mainly carbon that is still classed as a ferrous alloy. Carbon steel is an alloy of iron with up to 2% w/w carbon that increases the strength of the material and its corrosion resistance. The steel also contains trace quantities of other metals, such as nickel or chromium.

Nickel can be included in steel alloys to increase strength and corrosion resistance. *Nickel alloys* are commonly used in the *Christmas tree*—a combination of valves and piping that fit over an oil or a gas wellhead (Speight, 2014). Steel with a 9% w/w nickel content is durable at extremely high temperatures as well as at very low temperatures. It is used in heat exchangers, which remove heat from oil and gas at approximately 200°C (390°F) and cool it to 21°C (70°F), allowing the safe transportation of the fluid. Steel and nickel alloys are used extensively in gas processing plants and liquefied natural gas (LNG) plants because of their high strength and corrosion resistance.

Copper and *copper-containing alloys*, such as bronze, have excellent electrical and thermal conductivity and cryogenic, or cold-resistant, properties. These metals are used in valves, stems, seals, and heat transfer applications. A bronze alloy with traces of nickel and aluminum can be used in wellhead equipment and blowout prevention valves. Copper salts, such as copper sulfate, are used in gas processing plants to absorb mercury, which occurs in natural gas and separates out during processing—mercury is toxic to humans and reacts with other metals within the processing plant equipment and can cause brittle failure.

Titanium is one of the most versatile and valuable metals used in the oil and gas industry—the addition of titanium to steel alloys increases alloy strength, density, and corrosion resistance. The use of titanium-containing alloys in downhole tubing—as part of the steel alloy lining around a well bore—has become widespread over the past decades. High-strength titanium alloys used in compressor parts are durable and increase the useful lives of those parts compared with other steel alloys. In addition, titanium is highly resistant to seawater, carbon dioxide, and hydrogen sulfide corrosion and maintains alloy strength at the very low temperatures ($-150°C$, $-240°F$) required to liquefy natural gas. Titanium is also used in heat exchanger tubing in LNG plants and in the linings of the pressurized vessels in LNG tankers.

Chromium was one of the first metals to be used in strengthening steel alloys and is still used for this purpose. Low-carbon steel containing $12-14\%$ w/w chromium is highly resistant to carbon dioxide, hydrogen sulfide, and the high temperatures (in excess of $225°C$, $435°F$) found in the deep oil and gas wells and many process units. The use of steel tubing containing this proportion of chromium has soared over the last 5 years in the United States with the interest in drilling and recovering gas from tight shale formations. Chromium compounds such as chromium lignosulfonate have been used in oil industry drilling fluids as deflocculants to reduce the viscosity of the fluid and prevent suspended rock material from clogging around the drill bit. However, chromium compounds have been shown to damage the environment and have been replaced in drilling fluids by iron lignosulfonate and calcium lignosulfonate.

Molybdenum is another metal that increases the strength and corrosion resistance of steel alloys. High-performance steel used for gas pipeline construction contains from 2% to 4% w/w molybdenum. This metal is also used as a catalyst in refinery processes that remove sulfur to produce low-sulfur gasoline and other fuels that meet environmental regulations (Speight and Ozum, 2002; Hsu and Robinson, 2006; Gary et al., 2007; Speight, 2014). In fact, the typical alloys used in a refinery for elevated temperature service greater than 260°C (500°F) typically contain 0.5−9.0% w/w chromium plus molybdenum. Typically, at least 5% w/w chromium is required to resist oxidation at temperatures in excess of 430°C (800°F). Currently, most refineries use 9Cr−1Mo (an alloy containing 9% w/w chromium and 1% w/w molybdenum) tubes in coker heaters. For carbon steel and low-alloy steel, creep becomes a design consideration at about 430°C (800°F) and 480°C (900°F), respectively. These alloys are used for pressure vessels, piping, exchangers, and heater tubes.

Carbon steel is by far the most common structural material in refineries due primarily to a combination of strength, availability, relatively low cost, and a resistance to fire. The low-alloy steels are specified for applications that require higher properties than can be obtained with carbon steel, which is often used in reactors requiring temperatures on the order of 480−515°C (895−960°F). Most of the use is limited to temperatures in the range 315−345°C (600−655°F) due to loss of strength and susceptibility to oxidation, as well as other forms of corrosion at higher temperatures. A large percentage of the applications for stainless steel and heat-resistant materials above 650°C (1200°F) are used in connection with fired heaters.

Carbon steel is commonly used for temperatures up to approximately 260°C (500°F), and then materials such as 5% w/w chromium, 0.5% w/w molybdenum, and 9% w/w chromium with 1% w/w molybdenum are substituted to provide better resistance to a hydrogen sulfide attack. Even these materials suffer from heavy corrosion and metal wastage as temperature, pressure, and sulfide concentration are increased. When the severity of the corrosion process is deemed too aggressive for steel, the austenitic stainless steels and high-nickel alloys are specified.

2.2.2.2 Austenitic Alloys

Austenitic stainless steel (an alloy of iron, usually containing at least 8% w/w nickel and 18% w/w chromium, used where corrosion resistance, heat resistance, creep resistance, or nonmagnetic properties are required) and high-nickel alloys are used in refinery and petrochemical applications in which superior corrosion resistance and high-temperature strength are required. Materials intended to resist high-temperature hydrogen sulfide in refineries include a range from carbon steel to high-nickel alloys. Severe problem areas for this type of corrosion include applications such as heater tubes and transfer lines, piping exchangers, and air coolers.

The structure of austenitic stainless steel provides a combination of corrosion, oxidation, and sulfidation resistance with high creep resistance, toughness, and strength at temperatures greater than 565°C (1050°F). Such alloys are, therefore, often used in refineries for heater tubes and heater tube supports, fluid catalytic cracking units, catalytic hydrodesulfurization units, amine units, sulfur recovery units, and hydrogen plants. However, the alloys are susceptible to grain boundary chromium carbide precipitation sensitization when heated in the range of 540°C (1000°F) to 820°C (1500°F). Where sensitization is to be avoided, refineries prefer to use the stabilized grades of Type 347 stainless steel or Type 321 stainless steel.

The American Iron and Steel Institute (AISI) Type 347 stainless steel is an Austenitic Standard grade stainless steel, which is commonly called AISI Type 347 chromium–nickel steel. It is composed of (in weight percentage) 0.08% carbon (C), 2.00% manganese (Mn), 1.00% silicon (Si), 17.0–19.0% chromium (Cr), 9.0–13.0% nickel (Ni), 0.045% phosphorus (P), 0.03% sulfur (S), 0.7% niobium (Nb), 0.1% tantalum (Ta), and the base metal iron (Fe). Other designations of AISI Type 347 stainless steel include UNS S34700 and AISI 347.

AISI Type 321 (S32100) is a stabilized stainless steel that is resistant to intergranular corrosion following exposure to temperatures in the chromium carbide precipitation range from 425°C to 815°C (800–1500°F). Type 321 stainless steel offers higher creep and stress rupture properties than Type 304 and, particularly Type 304L, which might also be considered for exposures where sensitization and intergranular corrosion are concerns. This results in higher elevated temperature allowable stresses for this stabilized alloy for ASME Boiler and Pressure Vessel Code applications. The Type 321 alloy has a maximum use temperature of 815°C (1500°F) for code applications like Type 304, whereas T304L is limited to 425°C (800°F).

However, the susceptibility of the austenitic stainless steels to stress corrosion cracking limits their use and requires special precautions during operation and at downtime. At downtime, the precautions taken to prevent stress corrosion cracking are either alkaline washing with a dilute soda ash and low-chloride water solution and/or nitrogen blanketing. The austenitic stainless steels are used for corrosion resistance or resistance to high-temperature hydrogen or sulfide damage. Solid stainless steel vessels are rarely constructed. Strip-lined, stainless-clad, or lined vessels are found in hydrocracking and hydrotreating services. Austenitic stainless steels also find service as tubing in heat exchangers exposed to corrosive conditions. Other chromium—iron stainless steels with little or no nickel form crystallographic structures that are different from austenitic stainless steel. This stainless steel alloy contains less than 0.10% w/w carbon, 11—13% w/w chromium, and the balance of iron, along with a ferritic structure.

2.2.2.3 Ferritic Alloys

Ferritic alloys are based on ferrite (also known as α-ferrite, α-Fe, or alpha iron), which is pure iron with a cubic crystal structure and which gives steel and cast iron their magnetic properties. With additions consisting primarily of chromium (0.5—9% w/w) and molybdenum (0.5—1% w/w), ferritic alloys are most commonly used at temperatures up to 650°C (1200°F). The higher strength as well as the ability to resist oxidation and sulfidation and also, in some cases, resistance to a noncorrosive but destructive environment (e.g., hydrogen) result in ferrite-containing alloys being the material of choice. However, these alloys have inadequate corrosion resistance to many other elevated temperature environments (such as hydrogen and hydrogen—hydrogen sulfide environments) for which nickel—chromium—iron (Ni—Cr—Fe) alloys are preferred.

When carbon steel or low-alloy steel is not suitable for specific applications, the common choice is from the 18Cr—8Ni (18% w/w chromium—8% w/w nickel) austenitic group of stainless steel alloys, which, with the 18Cr—12Ni steel alloys, are favored for their corrosion resistance in many environments and their oxidation resistance at temperatures up to 815°C (1500°F). However, at temperatures higher than 650°C (1200°F), the failing strength of these alloys becomes a consideration and more heat-resistant alloys must often be used.

Ferritic stainless steels are straight-chromium 400 series grades that cannot be hardened by heat treatment and can be only moderately hardened by cold working. They usually have a chromium content of between 10% and 20% and lower carbon contents than the martensitic stainless steels. Ferritic stainless steels are magnetic and have good ductility and resistance to corrosion and oxidation. They are inherently stronger than carbon steel and are utilized in applications in which thinner materials and reduced weight are advantageous. They are often specified for use due to their superior corrosion resistance and resistance to scaling at elevated temperatures.

However, when the ferritic stainless alloys are modified, they may be hardened and become what is called *martensitic* by heat treatment. The ferritic and martensitic stainless steels are classified by the AISI as the 400 series. The most common alloys from this series found in refineries are stainless steel types 410, 410S, 405, and 430. A common stainless steel for trays and lining in crude service is Type 410 stainless steel.

Martensitic grades of stainless steel were developed in order to provide a group of stainless alloys that would be corrosion resistant and hardened by heat treating. The martensitic grades are chromium steels and do not contain nickel. They are magnetic and can be hardened by heat treating and are typically used where hardness, strength, and wear resistance are required. Type 410 alloy is the basic martensitic grade, containing the lowest alloy content of the three basic stainless steels (304, 430, and 410 alloys). It is a relatively low-cost, general-purpose, heat-treatable stainless steel and is used widely where corrosion is not severe. Typical applications include highly stressed parts needing the combination of strength and corrosion resistance, such as fasteners. Type 410S alloy contains lower amounts of carbon than Type 410 and offers improved welding properties but lower hardenability—it is a general purpose corrosion and heat-resistant chromium steel recommended for corrosion resisting applications. Type 431 contains increased chromium for greater corrosion resistance and good mechanical properties, and typical applications include high-strength parts, such as valves and pumps.

Type 410 martensitic stainless steel plays a specific role in the refining and petrochemical industry and is an ideal material to be used where high-strength, moderate-corrosion resistance and heat resistance are required. The alloy hardens when it is heat treated. Chromium accounts for 12% w/w of the overall chemical composition, giving the steel the corrosion-resistant properties of stainless steel. Type 410 alloy is also recognized for its strength and is suitable for petrochemical components that are subjected to repeat abrasion and wear. Type 410 alloy exhibits limited corrosion resistivity and, as a result, should be used only in environments containing weak or diluted acetic acid, naphtha, nitric acid, and sulfuric acid. It performs well in applications in which low hydrogen sulfide exists. In the annealed state, the 410 alloy is ductile, easily formed, and machinable, but presents limited suitability for welding—a postweld heat treatment is required to prevent hydrogen stress corrosion cracking. In the petroleum refining industry, the 410 stainless steel alloy is primarily used in the construction of pumps, valve trim, tray valves, and fractionation towers.

2.2.3 Other Alloys
The principal nonferrous alloys in refinery and natural gas processing equipment are the copper-based and copper–nickel alloys. However, the use of copper-based alloys in ammonia (NH_3) or ammonium salt (NH_4^+) environments should be avoided—there is a high potential for corrosion/destruction of the alloy structure.

Aluminum brass, containing 2% w/w aluminum, has been found to be somewhat more resistant to erosion in saltwater. Inhibition with arsenic is necessary to prevent de-zincification, as in the case of admiralty brass. The stronger naval brass is often selected as the tube sheet material when admiralty brass (in which a small amount of tin improves resistance to seawater) tubes are used in condensers. Generally, bronze is a tin alloy of copper, although the term has been widely used for other alloys, including some brass alloys. Cast brass or bronze alloys for valves and fittings are usually copper–tin–zinc compositions, plus lead for machinability. Aluminum bronzes are often used as tube sheet and channel material for exchangers with admiralty brass or titanium tubes exposed to cooling water.

The 70/30 copper–nickel alloy is used for exchanger tubes when better saltwater corrosion resistance than in aluminum brass is needed, or when high metal temperatures in water-cooled exchangers may cause de-zincification in brass. Monel is a nickel–copper alloy with 67% w/w nickel and 30% w/w copper, which has very good resistance to saltwater and, under nonoxidizing conditions, to acids such as hydrochloric acid and hydrofluoric acid.

Monel has a better high-temperature resistance to cooling water than 70/30 copper–nickel. In fact, Monel cladding and Monel trays are commonly specified at the top of crude towers to resist corrosion by hydrogen chloride vapor and where the temperature is below 205°C (400°F). At temperatures in excess of 205°C (400°F), nickel-based alloys are attacked by hydrogen sulfide. For high-temperature strength and/or corrosion resistance, several nickel-based alloys are used for special applications, such as expansion bellows in fluid catalytic cracking process units (Alloy 625), stems in flue gas butterfly valves (Alloy X 750), and springs exposed to high-temperature corrosive gases and liquids (Alloy X). Titanium has excellent resistance to seawater, and it is also used for tubing in crude tower overhead condensers.

2.3 SELECTION OF MATERIALS

The equipment and components that comprise a refinery or petrochemical plant are fabricated using durable alloys that can withstand the harsh conditions of the refining industry—stainless steel is one such material (White and Ehmke, 1991; Garverick, 1994). The selection of materials for refinery construction (Table 2.2) depends on the type of refinery, the type of crude oil handled, and the process unit (Table 2.1); the expected life in service for each vessel is also a factor. While a modern refinery can benefit from many years of experience in the selection of materials for the various processing units—unit-specific or refinery-specific— corrosion problems still occur (depending upon the nature of the crude feedstock, which has changed substantially in the past three decades).

Often these problems can be solved by the selection of a suitable alloy, such as one of the nickel alloys or by the development of a unique unit design to take full advantage of the performance characteristics offered by these high-performance alloy products. Wrought and cast heat-resistant alloys are required in numerous refinery and petrochemical applications because of the combination of aggressive environments and strength requirements. High levels of nickel and chromium provide alloys with the capability of fulfilling these requirements on an economical basis.

The key to the selection process has been the need to meet minimum performance criteria in terms of corrosion resistance at the lowest practical cost and with a minimum of fabrication constraints. However, the ever-increasing use of sour crude feedstocks and the push to greater-than-design production rates have resulted in an increasing frequency of operating problems with traditional materials. For example, excessive throughput of sour crude oil through a desalter could result in higher chloride concentrations entering the atmospheric distillation tower.

Table 2.2 Various Alloys and Their Respective Composition

Alloy Class	Example	Constituents									
		Ni	Cr	Mo	Fe	Co	Ti	Cu	Cb	Al	V
Carbon steel	C10				>94						
Low-alloy steel	1-1/4Cr 1/2Mo		1.25	0.5	Balance						
Fe—Ni—Cr + Mo	Type 316L	13.0	17.0	2.3	Balance						
	Alloy 800H	32.5	21.0		4.6						
	20Cb-3	35.0	20.0	2.5	Balance			3.5			
Ni—Cr—Mo	Alloy C2	54.0	15.5	16.0							
	Alloy C-276	57.0	16.0	16.0	5.5						
	Alloy C4	54.0	16.0	15.5	3.0						
	Alloy 625	60.0	21.5	9.0					3.7		
Ni—Cr—Fe	Alloy G	45.0	22.2	6.5	19.5			2.0			
	Alloy 600	76.0	15.0		8.0						
Ni—Mo	Alloy B2	Balance	1.0	28.0	2.0	1.0					
Ni—Cu	Alloy 400	65.1						32.0			
Nickel	Alloy 200	99.9									
Co-Base	ULTIMET (R)	9.0	26.0	5.0	3.0	54.0					
Ti-Base	Ti-6Al-4V						90			6.0	4.0

The chlorides hydrolyze to hydrogen chloride and, subsequently, corrode portions of the atmospheric tower, the overhead condenser, and the connecting transfer lines.

In most refinery and petrochemical applications, the unit components must be resistant to multiple forms of corrosion. A material that can resist general corrosion or pitting or chloride or polythionic acid stress corrosion cracking (PTA-SCC) is not adequate. Polythionic acid is formed in the presence of sulfur, moisture, and oxygen. Sulfur can come from feedstock, additives, or fuels. This failure mechanism is possible from both inside and outside of a tube, depending on the sulfur source. In fact, intergranular stress corrosion cracking is most common in sensitized 300 series stainless steel and higher nickel base austenitic alloys—alloys that are a nonmagnetic solid solution of ferric carbide or carbon in iron. Thus, nickel-base products alloyed with chromium and molybdenum and stabilized by either titanium or niobium are widely applied (Shoemaker et al., 2008).

Materials methods include selecting the proper material for the application. In areas of minimal corrosion, cheap materials are preferable, but when severe corrosion can occur, more expensive but longer lasting materials should be used. Other methods for mitigating corrosion come in the form of protective barriers between corrosive substances and the equipment metals. These can be either a lining of refractory material such as standard Portland cement or other special acid-resistant cements that are shot onto the inner surface of the vessel. Also available are thin overlays of more expensive metals that protect cheaper metal against corrosion without requiring lots of material.

Thus, the selection criteria for choosing a suitable metal alloy for use in refining equipment must include qualities that are either desirable or necessary—unfortunately, the optimum properties associated with each selection criteria seldom occur in a single material, especially when the operating conditions become aggressive, as is the case in the petroleum industry (recovery and refining). Compromises may be necessary to realize the best performance of the material selected. The principal selection criteria applied to materials for refining, gas processing, and petrochemical plant equipment include, but are not necessarily limited to: (1) mechanical properties, (2) corrosion resistance, (3) stability of the material, (4) ease of fabrication, and (5) availability and cost, both of which are interrelated.

2.3.1 Distillation

After the dewatering/desalting operation, distillation is the process used to separate crude oil into a variety of fractions. In addition, corrosion due to salts and renegade brine passed over from the desalting operation can cause major corrosion problems. Thus, various types of stainless steels are used in crude distillation units for protection against elevated temperature sulfidation, against attack by naphthenic acids, if present, as well as against attack by hydrogen chloride. Corrosion by naphthenic acid derivatives is severe in the 288−345°C (555−650°F) range with some decrease in corrosivity at temperatures above 400°C (750°F) at which the acids are believed to decarboxylate either thermally or catalytically (Zhang et al., 2005, 2006):

$$RCH_2CO_2H \rightarrow RCH_3 + CO_2$$

If this occurs in the distillation unit, the possibility arises of corrosion caused by carbon dioxide.

The stainless steels used most frequently in crude units are Types 304, 316, 405, and 410. Type 304 has excellent resistance from 260°C to 400°C (500−750°F). Type 316 is the standard molybdenum-bearing grade, second in importance to Type 304 among the austenitic stainless steels. The molybdenum gives Type 316 better overall corrosion-resistant properties than Type 304, particularly higher resistance to pitting and crevice corrosion in chloride environments. It has excellent forming and welding characteristics. Type 316L, the low-carbon version of 316, is immune from sensitization (grain boundary carbide precipitation). Type 316H, with its higher carbon content, has application at elevated temperatures—the austenitic structure also gives the Type 316 stainless steels excellent durability, even under cryogenic conditions as might be used in gas processing plants.

With naphthenic acid corrosion, there is little advantage to using the straight-chromium stainless steels. With low neutralization numbers, Type 304 has been used with some success, but is subject to corrosive wear, and the preferred alloy for naphthenic acid is Type 316. In crude oils containing naphthenic acids, sulfur is typically present, and Type 316 has excellent resistance to both. Type 329 (austenitic standard grade chromium−nickel steel) is a candidate for most corrosive applications in overhead condensers.

One factor to consider in the use of stainless steels—in addition to protection in sulfur and naphthenic acid environments—is the minimum scaling tendencies. Crude unit streams go to other refinery processes that have packed catalyst beds, such as desulfurization and hydrocracking. These beds can become plugged if there is scale or iron in the streams, resulting in downtime and lost production. Scaling is often not included in reactor and transfer line design corrosion rates, but scaling nevertheless occurs in other systems (including heat exchanger tubes and fractionation columns), even at low corrosion rates.

2.3.2 Coking

Coking processes are processes in which the feedstock is thermally decomposed into lower boiling products. The major subforms of the coking process are (1) the delayed coking process and (2) the fluid coking process.

To a large extent, most cokers are constructed of Type 410 stainless steel, either claded (i.e., bonded with another metal) or solid, except those with high naphthenic acid content, which would normally require Type 316. The Type 400 series with greater than 12% w/w chromium as well as the duplex stainless steels (which have a mixed microstructure of austenite and ferrite, the aim usually being to produce a 50/50 mix, although in commercial alloys the ratio may be 40/60) are subject to embrittlement when exposed to temperatures of 370−510°C (700−950°F) over an extended period of time. This is often referred to as 885F embrittlement (885°F = 475°C) because this is the temperature at which the embrittlement is the most rapid. This type of embrittlement results in low ductility and increased hardness and tensile strengths at room temperature, but the steel retains desirable mechanical properties at operating temperatures.

There is the potential for 885F embrittlement in coke drum overhead lines where temperatures may reach 430°C (810°F), but at lower temperatures, such as below 370°C (700°F), there does not appear to be any embrittlement, so Types 410 and 405 are perfectly satisfactory. If 885F embrittlement has occurred, replacement with Type 304 or 316 may be adequate.

Effluent from the coker is at a temperature on the order of 425°C (800°F), which requires 12% w/w chromium stainless steel or better. If the corrosion is aggressive, Type 304 or 316 will have to be used—Type 304 is the more likely candidate because there is a large amount of diluent in naphthenic acids. The flow recirculates between the fractionator, point 6, and the coke drums, with cooling provided partly by incoming feed and partly by the pan tray at 400°C (750°F). At this temperature, mitigation of corrosion by naphthenic acids requires Type 316 stainless steel—if naphthenic acids are absent, Type 410 is adequate. The gas oil stripper bottom, which is one of the products of the process, may require Type 410, 304, or 316 for the pump, piping, and hottest exchanger. Type 410 is generally favored because there is an effort to reduce the use of low-chromium steel, which tends to introduce dissolved iron (as iron naphthenate) and scale into the feed for the hydrocracker or desulfurizer leading to increased plugging of the reactor catalyst beds.

2.3.3 Fluid Catalytic Cracking

Fluid catalytic cracking is a post—WW II form of cracking in which catalysts aid the process. The fluid catalytic cracking process consists of two subunits—the reactor and the regenerator.

The stainless steels used in fluid catalytic cracking are mostly Types 304, 321, and 347, although there may be some applications for Types 405 and 410. The austenitic grades have excellent high-temperature strength characteristics and are resistant to oxidation and sulfidation. Old regenerators normally operate at 565°C (1050°F), but as newer catalyst types are used, temperatures may be as high as 760°C (1400°F). At these temperatures, carbon steel lacks sufficient strength and the chromium steels tend to carburize. Consideration must also be given to oxidation and sulfidation.

At the entry side of the reactor, the temperature normally runs 495–525°C (920–980°F), and this can be a sulfidizing environment. Carbon steel is used to a great extent, although there are places where stainless steels are used, such as the cyclone plenum and hanger rods (in both reactor and regenerator). The stainless steels used for hanger rods are Types 304, 321, or 347, with Type 321 being the most popular. The reactor liner and cyclones are frequently of 12% w/w chromium stainless steel, either Type 405 or 410. An important consideration is thermal expansion. A 12% w/w chromium stainless has about the same coefficient of thermal expansion as carbon steel.

The overhead line from the reactor is either carbon steel or Type 304, depending on the nature of the process and sulfur content of the feed. If sulfur content is high, stainless steel is usually preferred. In the fractionator, the lower half is lined because the operating temperature range is 370–480°C (700–900°F). These linings historically have been

Type 405 or 410, although some are of Type 304. The top of the fractionator is usually lined with a higher alloy material because of potential problems with ammonium chloride.

Many fractionators have water condensers in the overhead, and the materials used in these condensers need resistance to corrosives in the cooling water side and to ammonia, hydrogen chloride, and hydrogen sulfide in the overhead stream side. Brass has been used to some extent; but with high-nitrogen content feed, brass tends to pit. Also, sulfidation is a problem with brass and other copper alloys. Monel has been used where ammonia is not too high. Stainless steels are used, especially the duplex Type 329, which has good resistance to pitting and sulfidation. Also, because of high-nitrogen content, fluid catalytic crackers tend to produce cyanides, which increase corrosion rates. Stainless steels with a high molybdenum content are required if the cyanide levels are high.

2.3.4 Hydroprocesses

Hydroprocesses (hydrogenation processes) for the conversion of petroleum fractions and petroleum products are classified as destructive and nondestructive (Speight and Ozum, 2002; Hsu and Robinson, 2006; Gary et al., 2007; Speight, 2014).

Hydrotreating or *hydrodesulfurization* (*nondestructive hydrogenation*) is used for the purpose of improving product quality without appreciable alteration of the boiling range. Mild processing conditions are employed so that only the more unstable materials are attacked. Nitrogen, sulfur, and oxygen compounds undergo reaction with the hydrogen to remove ammonia, hydrogen sulfide, and water, respectively. Unstable compounds that might lead to the formation of gums or insoluble materials are converted to more stable compounds.

Hydrocracking (*destructive hydrogenation, hydrogenolysis*) is characterized by the conversion of the higher molecular weight constituents in a feedstock to lower boiling products. Such treatment requires severe processing conditions and the use of high hydrogen pressures to minimize polymerization and condensation reactions that lead to coke formation.

2.3.4.1 Hydrotreating and Hydrodesulfurization

Hydrotreating processes are catalytic processes using hydrogen to perform very mild hydrogenation for removal of sulfur and nitrogen from the feedstock. Thus, the corrosive substances to be formed in hydrodesulfurization units are hydrogen sulfide (in the presence of hydrogen at elevated temperature), ammonia and ammonium hydrosulfide (at low temperatures), and polythionic acids (formed from metal sulfides, air, and moisture during shutdowns). Ammonia chlorides are often present. Hydrogen at high temperature and pressure attacks the structure of ordinary steel causing embrittlement and reduction of mechanical properties.

Stainless steels are preferred for sulfur-containing environments over 290°C (550°F) and for the fact that they show less tendency to coke. The stabilized grades are used to prevent sensitization and minimize the possibility of PAT-SCC during shut down. The stainless steels used in hydrodesulfurization and hydrofining are Types 405, 410, 430, 304, 321, 347, and 308 for weld overlay. The choice of material, to a great extent, depends on the amounts of hydrogen sulfide and hydrogen, the temperature, and the high-temperature properties of the alloys. At low temperatures, when moisture is present, the choice revolves around chemical corrosion resistance, stress corrosion resistance, and corrosion erosion resistance.

Starting with the *feed-effluent* heat exchangers, when the temperature of any components (i.e., tubes, tube sheets, shell, and support baffles) is over 290°C (550°F), the most common type of stainless steel is Type 347, but Types 321, 304, and 430 have been used—the reason for stainless is to resist high-temperature sulfidation. The hot piping from the feed-effluent exchanger is fabricated from Types 347, 321, or 304 stainless steel. At this point, recycle and make-up hydrogen join the feed stream—hydrogen reduces the tendency for coking. Low-pressure reactors that operate at approximately 700 psi are clad with Types 347, 321, 304L, or 410, but high-pressure units are built like hydrocrackers with a more economical weld overlay, either Type 308 or 347. Bed supports, distributor trays, thermowells, and scale baskets inside the reactors are always stainless steel, usually Types 347, 321, or 304.

Polythionic acid will cause intergranular corrosion cracking or stress corrosion cracking of sensitized and mildly stressed austenitic stainless steel. Types 304 and 316 can be sensitized from welding, from vessel stress relief, from hot regeneration temperatures, and from long periods of time at relatively low temperatures, approximately 410–425°C (775–800°F). Polythionic acid does not normally cause corrosion to fully annealed, stabilized (nonsensitized) stainless steel. Failures due to polythionic acid intergranular cracking have been reported in piping, cladding, tube sheets, fired heater tubes, and heat exchanger tubing.

The trays in the fractionator are normally Type 410, and on occasion Type 304. The gas cooler and naphtha coolers are candidates for 304, 316, 18-2, or 26-1 if long life and cleanliness are desired in the overhead system. Stainless steels 304 and 18-2 would be used for modestly aggressive waters, while 316 and 26-1 would be better suited to high-chloride cooling waters.

2.3.4.2 Hydrocracking

Hydrocracking is a combined desulfurization and cracking operation that can convert a full range of hydrocarbon feedstocks into more valuable products. The conversions occur in the presence of high-pressure hydrogen; hence hydrocracking is hydrogenation or addition of hydrogen to molecules. The corrosion problem considerations and remedies are quite similar to those in hydrotreating.

The stainless steels used in hydrocracking desulfurizing reaction section are Types 304, 321, 347, 410, and 430. Austenitic steel is preferred because of the superior high-temperature strength, and because of potential problems of 885F embrittlement with ferritic stainless steels. However, ferritic stainless steels, Types 430 and 26-1, have been used as the tube material for feed-effluent exchangers operating below 375°C (705°F). The newer grade 18-2 would be satisfactory in the same locations. The stabilized grades are used to prevent sensitization—especially in welds—and the possibility of polythionic stress corrosion cracking.

With the feed and recycle gas flowing through heat exchangers, it is safe to assume that the stream temperature will be above 290°C (555°F) and the use of stainless steel is justified. The reactor effluent on the opposite side is certainly above 345°C (650°F), so regardless of what materials are used on the feedstock side of the reactor, the effluent requires stainless steel. Type 304 tubing has been used, but Types 321 and 347 stainless steels are preferred because of lower susceptibility to polythionic stress corrosion cracking.

2.3.5 Other Processes

The processes briefly described above do not produce saleable products. The products of these processes have to be treated further to meeting specifications and to be ready for the market. A variety of other processes (often called *product treating processes*) are available for such treatment (Speight and Ozum, 2002; Hsu and Robinson, 2006; Gary et al., 2007; Speight, 2014).

2.3.5.1 Catalytic Reforming

Catalytic reforming increases the antiknock quality of motor fuel blending stocks. The principal reaction in this process is dehydrogenation of naphthenes to form aromatics. As a result, hydrogen is produced, some of which is recycled to sustain reformer reactor pressure and to reduce coke formation. Most of the hydrogen is available for hydrotreating, hydrocracking, and other processes. The feedstock for catalytic reforming is naphtha, which has been desulfurized. If the sulfur compounds are not removed, poisoning of the catalyst is guaranteed. Because sulfur is removed, the stream is not very corrosive, and carbon or low-chromium steels are adequate for most applications— except possibly in the final reformate cooler. Also, reactor internals, such as thermowells, catalyst support devices, and screens are usually Type 304, which has the high-temperature strength.

There are two potential corrosion problems in the reformate cooler: (1) from the cooling water and (2) due to the use of organic chloride to regenerate the catalyst. The organic chloride is added to the stream continuously to keep the catalyst active. When the stream gets to the final cooler, the temperature is reduced to the point at which ammonium chloride can precipitate, and if water is present, this can be very corrosive to steel and brass. Moist ammonium chloride will cause pitting of Types 304 and 316. Consequently, if stainless steel is picked for this component, it must be adequate to resist pitting on the cooling water side. Stainless alloys best suited for the final reformate cooler are Types 329, 6X (an austenitic nonstandard grade stainless steel), or 29-4 (an austenitic nonstandard grade stainless steel). The use of Type 304 trays, downcomers, and beams in the stabilizer reduces iron corrosion-product tray fouling.

2.3.5.2 Alkylation

Alkylation is defined broadly as combining an olefin with an aromatic or a paraffin hydrocarbon using a catalyst—the most common catalyst is sulfuric acid.

Alkylation plants are built basically of carbon steel, but sulfuric acid–resistant alloys are required in some areas. The alloys are primarily Type 316, 316L, 20Cb-3, and nickel-base alloys. Although carbon steel is resistant to concentrated sulfuric acid in the strength range of 90–98% found in alkylation, steel loses resistance under high-velocity conditions and above 38°C (100°F). At 90% acid concentration, the corrosion rate is sufficiently low and may not justify the use of alloy materials. However, if acid dilution is anticipated, which will cause a rise in temperature, carbon steel is not effective because corrosion rates increase rapidly over 38°C (100°F).

Downstream of the water wash system, traces of acid and water in corrosive dilute concentration occur from entrainment or carryover. Also, sulfate–hydrocarbon compounds are formed in the reaction which breaks down with heat, liberating sulfur dioxide. Spent alkylation acid contains an appreciable concentration of entrained light hydrocarbon that must be removed by flashing, which is corrosive to steel, or by a slow weathering process. Alloy 20Cb-3 is frequently used in the flash equipment for heat exchanger tubing, flasher overhead piping, as well as the lining and internals of the flash drum.

2.3.5.3 Polymerization

Polymerization in the petroleum industry is the process of converting olefin gases including ethylene, propylene, and butylene into hydrocarbons of higher molecular weight and higher octane number that can be used as gasoline blending stocks. The olefin feedstock is pretreated to remove sulfur and other undesirable compounds. In the catalytic process, the feedstock is either passed over a solid phosphoric acid catalyst or comes in contact with liquid phosphoric acid, where an exothermic polymeric reaction occurs. This reaction requires cooling water and the injection of cold feedstock into the reactor to control temperatures between 150°C and 235°C (300°F and 450°F) at pressures from 200 to 1200 psi. The reaction products leaving the reactor are sent to stabilization and/or fractionator systems to separate saturated and unreacted gases from the polymer gasoline product.

The potential for an uncontrolled exothermic reaction exists should loss of cooling water occur. Severe corrosion leading to equipment failure will occur should water make contact with the phosphoric acid, such as during water washing at shutdowns. Corrosion may also occur in piping manifolds, reboilers, heat exchangers, and other locations where acid may settle out.

Phosphoric acid is generally less corrosive to many metals and alloys than are sulfuric and hydrochloric acids. For several nickel-containing alloys, it is only mildly corrosive at temperatures below about 100°C (212°), independent of concentration. Beyond this temperature its corrosivity becomes substantial toward most metals, regardless of whether they are normally considered resistant to oxidizing or reducing conditions. The properties of the acid are affected quite substantially by its process of manufacture and its impurities. Most nickel alloys that have been reported or used to some extent in phosphoric acid processes are wrought alloys, used largely in the form of sheet, plate, pipe, tubing, or bar.

2.3.5.4 Hydrogen Production
Hydrogen for hydrotreating and hydrocracking processes is produced by the classic reaction of steam with a hydrocarbon (steam reforming) at a high temperature over a catalyst.

The stainless steels used most frequently in hydrogen plants are Types 304, 304L, 310, 330, 410, 430, and the cast grade Type ACI HK-40. The environments of principal concern for stainless use are high temperature, high velocity, steam, steam condensates very rich in carbon dioxide (carbonic acid), and the need to keep steam condensate free of iron contamination. Since the nickel catalyst is poisoned by sulfur, the feed is desulfurized in various ways, including combinations of activated carbon, amine scrubbing, caustic scrubbing, and catalytic desulfurization, followed by zinc oxide absorption. Since a great deal of stainless steel is utilized and is necessary, steam and process water must be kept pure for process reasons and free of chloride to avoid stress corrosion cracking.

Carbon steel is very actively corroded by strong carbonic acid. Chromium additions to steel reduce the rate of attack. Steels that contain at least 12% w/w chromium are found to be extremely resistant to carbonic acid for use in cooled shift gas, carbon dioxide absorption, and solvent regeneration. The vertical reformer, catalyst-filled tubing, with skin temperature from 705°C (1300°F) at the top to 955°C (1750°F) at the bottom, is predominately ACI HK-40 alloy. Several of the early version low-pressure reformers used Alloy 800, which is similar to Type 330 stainless steel.

From the steam generator through feed-versus-reformer-effluent exchangers and through the shift converters, there is no requirement for stainless steel. But when the shift gas contains substantial amounts of with carbon dioxide, and steam drops in temperature below the dew point, a corrosion-resistant alloy is required—Type 304 is the common choice although 430, 410, 18-2, 26-1, or any stainless steel is satisfactory.

When the shift gas contacts the carbon dioxide removal solvent, such as monoethanolamine (MEA) or potassium carbonate, the carbon dioxide is captured and inhibited as long as it is held in solution. The absorber can be built of steel, although screens for packing support might be Type 304. In the upper portion of the absorber, condensate is injected onto several wash trays to remove traces of the solvent, to prevent carryover into the methanator. The trays and tower shell in this area should be Type 304. Both Type 304 and 304L alloys have been used for piping and tower linings where fusion welding is required and carbide precipitation will occur. If there is any doubt about intergranular corrosion in a particular plant, corrosion testing of sensitized specimens might be the best course of action.

If the purified hydrogen is relatively free of moisture and carbon dioxide, there is no need for stainless steel piping, beyond the absorber or beyond an overhead knockout pot. The gas to the methanator preheater, the methanator, and to refinery hydrotreat is noncorrosive.

2.3.6 Sour Water Strippers

Sour water strippers are used to remove pollutants from various refinery wastewater streams. Pollutants include hydrogen sulfide, ammonia, carbon monoxide, carbon dioxide, cyanides, thiocyanates, phenols, salts, organic and inorganic acids, inhibitors, and some hydrocarbons.

In a basic stripper unit, steam, injected at the bottom, carries the gaseous pollutants overhead for further treatment. Overhead vapor is condensed and returned to the tower. Effluent water, which is relatively clean, is reused in the refinery or sent to final treatment in a refinery safety basin system. Heat exchangers are used to heat feed, reboil effluent to produce steam, condense overhead, and to recover heat. The type and quantity of pollutants that have to be removed depend on the type of crudes handled in the refinery and the types of processes used. Consequently, sour water strippers encounter a wide variety of corrosive environments. However, the most common problem is sulfide corrosion from hydrogen sulfide, which is increased by the presence of cyanides.

There are three basic stripper types: (1) acidified, (2) nonacidified noncondensing, and (3) nonacidified condensing. There are only a few acidified units, the rest are divided evenly between condensing and noncondensing strippers that use straight sour water feed without acid addition.

In acidified units, there are applications for stainless steel primarily in the tower and the acid mixing column. By adding sulfuric acid to tie up ammonia, stripping of hydrogen sulfide is much more complete. Type 316 has adequate resistance to 20% sulfuric acid at low temperature, so it can be used in feed lines, tower, and trays. Alloy 20Cb-3 has been used for tower lining and trays also. In the mixing column, however, where acid concentration is higher, it would be more desirable to use higher alloyed material, such as 20Cb-3.

The overhead exchanger is either a tube-type unit or air cooler in which there may be plugging from ammonia bisulfide. As plugging occurs, velocities increase with a resulting increase in corrosion. Type 304 or 316 has been used in these units to minimize the problem.

Reflux drums are usually carbon steel. There is the potential for hydrogen blistering and cracking welds, but these problems can be controlled by controlling cyanides. Reflux pumps, however, usually require Type 20Cb-3 or a high-nickel alloy. If metallic control valves are used, these can be Type 304 or Type 316. The effluent cooler is another stainless application. Although many coolers have been constructed of carbon steel, failures have been reported. Several tests indicate that Type 304 or 316 is adequate, remaining clean and not showing signs of deposits and pitting.

A nonacidified system is more alkaline, but this tends to drive the cyanides overhead along with ammonia and hydrogen sulfide. The tower can be carbon steel with a liberal corrosion allowance, but the trays should be stainless, either Type 410, 304, or 316. The overhead system, however, is very aggressive and high-alloy materials should be used, such as Type 6X, 29-4-2, 29-4, or titanium. The reflux lines in a nonacidified unit can be Type 304 or 316, and the bottom of the reflux drum can be lined with stainless steel.

Reboilers to generate heat and steam for stripping are occasionally used in sour water stripping plants in place of direct steam introduction. Reboilers are used both for acidified and nonacidified units. Reboiler service is very corrosive because polluted, salt-containing dirty water is being boiled. The environment is conducive to chloride stress corrosion cracking and pitting under deposits on the tubes. Reboiler shells are lined with Type 316, Type 20Cb-3, or, perhaps, a more corrosion-resistant alloy. Tubing candidate materials for reboilers are Type 6X, Type Nitronic 50, Type 29-4, or Type 29-4-2.

2.3.7 Cooling Water Systems

2.3.7.1 Freshwater and Polluted Water

Typically when freshwater is used for a cooling system, clean stainless steel heat exchanger tubing resists all forms of corrosive attack. The total amount of stainless steel tubing that has failed in electric utility cooling condensers, for example, including saltwater applications, is approximately one half of one percent ($<0.5\%$), which is exceptionally low considering the many millions of feet of tubing in service.

There are no satisfactory guidelines on the ability of stainless steels to stand up in polluted water because of the variety and nature of pollutants that can be involved. If the pollutants are in the form of sewage, stainless steels can be used because of the resistance to hydrogen sulfide, ammonia, and carbon dioxide. They are used extensively in sewage treatment plants. If the pollutants are effluents containing chemical wastes, which are becoming more unlikely because of tightening pollution control regulations, in all probability, chemical wastes will be sufficiently diluted to permit the use of stainless steels.

2.3.7.2 Polluted Saltwater

Where polluted saltwater has been used for cooling, stainless steels have demonstrated greater reliability than copper-base alloys under circumstances of abnormally high sulfide content. Sulfide attack of copper-base alloys is sufficiently rapid and certain to make failures from this cause far more important than an occasional failure of stainless steel by chloride-induced pitting. Furthermore, the danger of chloride pitting can be minimized in brackish water by using Type 316, Type 317, or one of the higher alloyed stainless steels—and by avoiding or preventing the accumulation of localized deposits that are potential pitting sites. For example, Type 6X and Nitronic 50 have been proven effective in severely polluted brackish water. In addition, ferritic stainless steels, specifically Types 29-4 and 29-4-2, are resistant to polluted or nonpolluted brackish water.

Stainless steel in seawater is a separate issue—some reports indicate that there has been success with Type 316, but only with continuous or periodic cleaning to prevent fouling. Spotty or continuous fouling in seawater leads to failure by pitting. An alternative is one of the higher alloyed stainless steels, which may be more economically attractive than titanium.

2.3.7.3 Cooling Towers

Water conservation has resulted in many refineries using cooling towers. From the standpoint of potential corrosion problems, towers can be both an asset and a liability. Because cooling tower water is recirculated, the plant can exercise some control over water quality by the use of pH control, chemicals to prevent scale, and corrosion inhibitors. With once-through cooling, such control is out of the question because of the costly treatment of huge quantities of water and newly enforced water pollution regulations. On the other hand, cooling towers concentrate the total dissolved salts occurring in the feed water, including chlorides.

Both Types 304 and 316 have been used successfully in cooling tower water situations when the chloride level is relatively low (500 ppm) and the exchangers remain relatively clean, or when corrosion inhibitors have been used in the water treatment program to protect steel heat exchanger tubing and steel piping. If chromate-base corrosion inhibitors cannot be used, it may be required to upgrade from Type 304 to Type 316, or to Type 317, or to one of the proprietary stainless alloys to avoid pitting corrosion under deposits.

In summary, stainless steels have been used successfully in virtually all cooling waters. For most cases below 500 ppm chlorides, Type 304 may be used, but for seawater, higher alloyed stainless steels containing molybdenum are a better choice. As water conservation regulations dictate more cycles of concentration and cutbacks of traditional corrosion inhibitors, cooling tower waters are expected to be more troublesome. However, stainless steels will continue to be used with reliance on the higher molybdenum grades, and, as for any material, cleanliness insures improved heat exchange and reduces risk of localized corrosion.

2.3.8 Gas Processing

Gas processing (often called *gas cleaning* or *gas refining*) consists of separating all of the various hydrocarbons and fluids from a variety of gas streams (natural gas, refinery gas, process gas) from the pure natural gas (Mokhatab et al., 2006; Speight, 2007). Gas processing is necessary to ensure that the gas intended for end use has the necessary composition. Gas processing is, in many respects, less complicated than the crude oil refining, but it is equally as necessary to remove contaminants and corrosive constituents.

2.3.8.1 General

Reactors in gas processing plants are built largely of carbon steel. However, there are components and areas where corrosion of steel is too high and stainless steel is more economical. Austenitic stainless steels, owing to their very low corrosion rates and their ability to withstand hydrogen sulfide without damage, make good liner materials for columns, and they provide long life for trays. The stripping section of the de-ethanizer is an area where blistering frequently occurs and Type 316 is used for the lining and internals.

Stainless steels are also used for resistance to general or uniform corrosion by acidic wet sulfur compounds, ammonium chloride, and cyanide that has not reacted with the inhibitor. Commonly used types are 304, 304L, 316, 316L, 329, 410, 405, and 330. In addition, Types 26-1, 6X, Nitronic 50, and 29-4-2 also find application, particularly when cyanides are causing serious corrosion.

Reboilers are low points in a fractionation system, which tend to trap and collect scale and other solids, to drop out water of absorption upon heating, and to collect entrained water. Carbon steel tubing frequently has a short life. The de-ethanizer and the depentanizer reboilers are subject to corrosion and pitting from hot corrosive water trapped in sludge both on the shell and tubes. The lower quadrant of reboiler shells are frequently lined with Type 316, while tubing is replaced with Type 329, 330, 20Cb-3, or more corrosion-resistant alloys. Corrosion of reboiler tubing poses a difficult problem because conditions exist for general corrosion, pitting, and chloride stress corrosion cracking of susceptible austenitic alloys. Highly corrosion-resistant austenitic and ferritic grades such as 6X, Nitronic 50, 29-4, and 29-4-2 may be appropriate.

The depentanizer preheater has shown corrosion from wet feed streams, so crack-resistant and corrosion-resistant Types 329, 330, and 26-1 have been selected for this service. Depentanizer columns have been reported to require Types 410, 304, or 316 internals to realize economical life. Type 410 tray components in the top portions of the depropanizer and debutanizer also have value.

2.3.8.2 Amine Plant

An amine system (*olamine* system) removes hydrogen sulfide and carbon dioxide from various fuel gases produced in a refinery, so as to prevent air pollution when the gases are burned. Amine plants consist basically of two units (Figure 2.1): (1) the absorber in which the unwanted constituents are removed and (2) the regenerator in which the solvent solution is reactivated. In the absorber, gas flows up through the column and the amine cascades down, providing intimate contact. Sweet gas exits the top and rich amine leaves through the bottom. In the regenerator, the hydrogen sulfide and carbon dioxide are driven off by heat and stripping steam. The lean amine solution is returned to the absorber.

In modern gas processing plants, the rich/lean exchanger train is often constructed initially of carbon steel. The corrosion problem depends

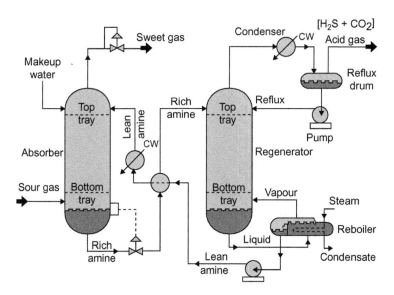

Figure 2.1 Schematic of an olamine unit (amine unit) for gas cleaning.

largely on the concentration of hydrogen sulfide. As with most plants, there is a tendency to build the unit too small, and, as it gets older, its capacity is increased. Thus, the hydrogen sulfide content of the rich amine solution increases and the corrosion problems get worse. The original carbon steel is then frequently replaced with stainless, beginning with Type 410 and then increasing to Types 430 and 304. The regenerator reboiler usually requires stainless, either Type 430 or 304, because of the higher temperatures involved.

Diethanolamine (DEA) is reconstituted using a precoat filter and treating heat-stable salts of DEA with soda ash. The filter elements are constructed of Type 304. MEA is reconstituted via a reclaimer which is in essence a miniature reboiler for driving off only steam and MEA vapor. Left behind for occasional blowdown or washing out are the heat-stable salts or polymerization products of MEA. Reclaimer service is *severe* on metals, and reclaimer tubing is usually 316, 329, or 20Cb-3.

In MEA plants, the corrosion problems appear to be more severe than in DEA plants. While carbon steel might be used in MEA, it is more likely that modern units will start with stainless steel in the rich/ lean exchanger train and in reboilers. In MEA plants removing *only* carbon dioxide, more severe corrosion problems can be expected. This is due largely to the fact that carbon dioxide is a stronger acid than hydrogen sulfide. When other solvents are used, such as Catacarb (typically hot potassium carbonate that may or may not contain a catalyst) and Sulfinol (a mixture of sulfolane, diisopropanolamine or methyl DEA, and water), consideration should be given to using Type 304 for the rich-lean exchangers, reboiler, and the overhead gas cooler.

2.4 PIPES AND PIPELINES

The selection of piping materials as pipelines for the transport of crude oil feedstocks to the refinery and products from the refinery is, in itself, a major issue in terms of the potential for corrosion and typically involves more than determining if a material is compatible with a given environment. Often adequate life can be obtained in corrosion services with carbon steel piping in conjunction with control of process and operating variables. In other cases, in particular those piping systems handling corrosive fluids at elevated temperatures, there is no alternative to corrosion-resistant materials. Also, low or elevated temperature service conditions can dictate the use of special materials.

Corrosion results from impurities in the hydrocarbon such as chloride salts, organic acids, water, and sulfur compounds or from by-products formed from the breakdown of these impurities. Also, chemicals added to hydrocarbons during processing, such as sodium hydroxide and sulfuric acid, may require the use of special metals and/or certain precautions.

The pipe specifications (ASTM A53 Grade B or ASTM A106 Grade B) for general hydrocarbon service (hydrocarbon services where corrosion is not expected and special requirements are not needed) are available for seamless pipe (*black iron pipe*). The basic specification for petroleum refinery service will require that valves have cast steel bodies with stainless steel trim, usually 12% w/w chromium stainless steel. Specifications for less severe service may allow cast iron flanged valves under the limits for ductile cast iron and for gray cast iron. Pipe is widely available and low in cost, can be bent hot and cold, and can be cut and welded using simple methods and minimal precautions. Carbon steel pipe has relatively high strength and ductility, adequate toughness for most applications, and fair resistance to corrosion in a wide range of environments.

2.4.1 Low- and High-Temperature Services
The fracture toughness of carbon steel and ferritic alloys decreases with decreasing metal temperature. Some ferritic materials, such as structural grade steels without chemistry limits and ductile and malleable iron, cannot be used below this temperature, but most ferritic steels can be used to a lower temperature provided they are stress relieved and qualified by impact testing.

Austenitic grades of stainless steel, provided they are in the solution-treated condition and contain less than 0.10% carbon, can be used to temperatures down to $-198°C$ ($-325°F$) without being impact tested. LNG, like other refrigerated hydrocarbons, is often handled in austenitic stainless steel pipe. Since austenitic stainless steel can be acquired as available (*off the shelf*) and applied directly to low-temperature service without special tests, there is a temptation to employ it automatically for temperatures under $-29°C$ ($-20°F$)—this may lead to unexpected problems, such as the occurrence of chloride stress corrosion cracking.

At elevated temperature, iron reacts chemically with elemental sulfur and/or sulfur compounds to form iron sulfide. The corrosiveness of the sulfur-bearing hydrocarbons, unlike chemical mixtures, is not proportional to the weight percent sulfur. The reason for this is that the sulfur may be present in various forms such as elemental sulfur, hydrogen sulfide, aliphatic sulfides, aromatic sulfides, polysulfides, mercaptans, and disulfides, all with different potentials for causing corrosion. Sulfide corrosion is strongly temperature dependent—at elevated temperatures many organic sulfides break down to form hydrogen sulfide or sulfur, which reacts with metal surfaces.

The sulfidation rate decreases in proportion to the amount of chromium in the steel. In crude fractionation units, carbon steel is relatively unaffected by corrosion at temperatures below 260−290°C (500−550°F) and marginal in performance at temperatures between 290°C and 345°C (550°F and 650°F). The most common carbon-to-alloy steel break temperature is approximately 290°C (555°F), but in some cases the use of alloy steel is required at temperatures as low as 260°C (500°F), while carbon steel has been used up to 315°C (600°F). When carbon steel is used in contact with sulfur at temperatures in excess of 260°C (500°F), it is common to specify silicon-killed grades such as ASTM A106 pipe and A 105 fittings. Steel with 0.15−0.30% w/w silicon has been shown to be greatly superior to steel with under 0.1% silicon in some environments.

2.4.2 Water−Hydrogen Sulfide Service

Another service condition calling for a separate specification is piping for either water or wet gas containing hydrogen sulfide. Carbon steel with extra corrosion allowance is usually suitable on the basis of metal loss, but some consideration must be given to the hydrogen that is charged into the steel due to corrosion in the presence of sulfide ions.

The primary consideration for sour service should be avoidance of hard valve components to avoid sulfide stress cracking. Sulfide stress cracking of valve components can have serious consequences, especially when it involves the valve stem. Not only is there a chance of leakage, but an open gate valve can fail closed and shut off a line. For this reason, it is good practice to make all process valves inherently stainless steel carbon (SSC) resistant (NACE Standard MR0175-90).

2.4.3 Hydrocarbon–Hydrogen Service

Hydrogen at high temperature and high pressure can permeate steel, and when the conditions are severe enough, react with metal carbides in the microstructure. Two types of damage are possible: (1) surface decarburization, which may not be serious, and (2) subsurface decarburization, which results in internal fissures that make the steel unsuitable for safe operation. Alloy steels containing chromium and/or molybdenum contain carbides more resistant to reduction by hydrogen.

Hydrotreating reactor inlet–outlet piping involves exposure of steels to hydrogen sulfide in the presence of hydrogen. There are various types of hydrotreaters, which is a general term to describe the catalytic desulfurization, treating, or cracking of hydrocarbons with hydrogen. All of the processes are similar and operate with reaction temperatures of approximately 370–455°C (700–850°F). The operating pressures vary from 400 psi for units designed to desulfurize light hydrocarbon streams to in excess of over 2500 psi in hydrocracking units designed to convert high-molecular-weight feedstocks to more valuable, low-boiling hydrocarbons. The piping for these two units may contain similar amounts of hydrogen sulfide, but the pipe materials may differ.

The most frequently used material for high-temperature hydrotreater piping is austenitic stainless steel, usually the titanium-stabilized Type 321 stainless steel. Austenitic stainless steels are not susceptible to embrittlement and have excellent ductility and toughness, even after long-term service. Austenitic stainless steels are susceptible to stress corrosion cracking when exposed to chloride environments.

The reactor circuit piping in hydroprocessing units is critical to their successful operation. Appropriate material selection is crucial because of the severe operating conditions that include high-pressure hydrogen, hydrogen sulfide, chlorides, nitrogen compounds, and other corrosives at temperatures in the range of 315–455°C (600–850°F). These conditions dictate the use of stainless steels (Singh and Bereczky, 2004). Thus, hydrocrackers and heavy gas oil desulfurization reactors present a more limited choice of piping materials than naphtha desulfurizers. 9Cr steel is not acceptable and while 12Cr stainless steel has an acceptably low corrosion rate, its low code stress values make it less attractive than austenitic grades of stainless steel. Also, its low toughness becomes more significant as the thickness of the pipe increases.

For hydrocrackers, a more economical alternative to extruded heavy wall Type 321 stainless steel pipe is centrifugally cast HF-modified austenitic–ferritic stainless steel piping, which is a casting alloy developed for this application. The alloy contains more carbon than wrought 18-8 grades of austenitic stainless steel, which makes the metal more fluid at casting temperatures and improves quality. Also, it is chemically balanced to produce a two-phase ferritic–austenitic microstructure, which ensures the production of sound, crack-free castings. The high chromium content gives the alloy very high resistance to high-temperature sulfide corrosion, but it causes the alloy to lose toughness after elevated temperature service. The loss of toughness is kept to within acceptable levels by controlling the ferrite level to under 15% w/w. The usual composition of HF-modified steel is chromium 21–25% w/w, nickel 6.5–11% w/w, carbon 0.15–0.20% w/w, and ferrite 5–15% w/w.

REFERENCES

ASME B31.3, 2012. Standard for Process Piping. Chemical Plant and Petroleum Refinery Piping Code. Standard A 53 Grade B. American Society of Mechanical Engineers (ASME), New York, NY.

ASTM A53/A53M, 2012. Standard Specification for Pipe, Steel, Black and Hot-Dipped, Zinc-Coated, Welded and Seamless. Annual Book of Standards. ASTM International, West Conshohocken, PA.

ASTM A106 Grade B, 2012. Standard Specification for Seamless Carbon Steel Pipe for High-Temperature Service. Annual Book of Standards. ASTM International, West Conshohocken, PA.

Fichte, R., 2005. Ferroalloys. Ullmann's Encyclopedia of Industrial Chemistry. Wiley-VCH Publishers, Weinheim, Germany.

Garverick, L., 1994. Corrosion in the Petrochemicals Industry. ASM International, Materials Park, OH.

Gary, J.G., Handwerk, G.E., Kaiser, M.J., 2007. Petroleum Refining: Technology and Economics, fifth ed. CRC Press, Taylor & Francis Group, Boca Raton, FL.

Hsu, C.S., Robinson, P.R. (Eds.), 2006. Practical Advances in Petroleum Processing Volume 1 and Volume 2. Springer Science, New York, NY.

Mokhatab, S., Poe, W.A., Speight, J.G., 2006. Handbook of Natural Gas Transmission and Processing. Elsevier, Amsterdam, the Netherlands.

NACE Standard MR0175/ISO15156. Petroleum and Natural Gas Industries—Materials for Use in H₂S-containing Environments in Oil and Gas Production. NACE International, Houston, TX, December 2005.

Shoemaker, L.E., Smith, G.D., Baker, B.A., Kiser, S.D., 2008. Resisting petroleum refinery corrosion with nickel alloys. Proceedings of CORROSION 2008. March 16–20, New Orleans, LA.

Singh, A.K., Bereczky, E.L., 2004. Toughness considerations for centrifugally cast HF-modified alloy piping in hydroprocessing services. Paper No. 04641. Proceedings of CORROSION 2004. NACE International, Houston TX.

Speight, J.G., 2007. Natural Gas: A Basic Handbook. GPC Books, Gulf Publishing Company, Houston, TX.

Speight, J.G., 2014. The Chemistry and Technology of Petroleum, fifth ed. CRC Press, Taylor & Francis Group, Boca Raton, FL.

Speight, J.G., Ozum, B., 2002. Petroleum Refining Processes. Marcel Dekker Inc., New York, NY.

Tillack, D.J., Guthrie, J.E., 1998. Wrought and Cast Heat-resistant Stainless Steels and Nickel Alloys for the Refining and Petrochemical Industries. Nickel Development Institute, Durham, NC.

White R.A., Ehmke E.F., 1991. Materials selection for refineries and associated facilities. Proceedings of CORROSION/91. NACE International, Houston, TX.

Zhang, A., Ma, Q., Wang, K., Tang, Y., Goddard, W.A., 2005. Improved processes to remove naphthenic acids. Final Technical Report, DOE Contract No. DE-FC26-02NT15383. US Department of Energy, Washington, DC.

Zhang, A., Ma, W., Wang, K., Liu, A., Chuler, P., Tang, Y., 2006. Naphthenic acid removal from crude oil through catalytic decarboxylation on magnesium oxide. Appl. Catal. A: Gen. 303 (1), 103–109.

CHAPTER 3

Corrosion in Refinery Units

3.1 INTRODUCTION

Petroleum refining is the separation of petroleum into fractions and the subsequent treating of these fractions to yield marketable products through the use of a series of unit processes in which each unit process carries out a separate function (Chapter 2) Refinery processes must be selected and products manufactured to give a balanced operation in which the petroleum feedstock is converted into a variety of products in amounts that are in accord with the demand for each (Speight and Ozum, 2002; Hsu and Robinson, 2006; Gary et al., 2007; Speight, 2014). Thus, a refinery is assembled as a group of integrated manufacturing plants that vary in number with the variety of products produced (Figure 3.1). Corrosion causes the failure of equipment items as well as dictating the maintenance schedule of the refinery, during which part or the entire refinery must be shut down (Kane, 2006; Dettman et al., 2010). Although significant progress in understanding corrosion has been made, it is also clear that the problem continues to exist and will become progressively worse through the introduction into refineries of more heavy crude oils, opportunity crudes, and high acid crudes.

The issue with corrosion in the refinery is that there is not one single source of corrosion, but many (Tebbal and Kane, 1996, 1998; Tebbal et al., 1997; Tebbal, 1999; Ayello et al., 2011). To add to the problem, some of the corrosion agents might be interactive and increase the corrosivity of each agent. Also, physical process conditions play a role: temperature and flow have to be taken into account, too, and of no less importance is the refinery infrastructure itself insofar as pipes, vessels, welding joints, and instruments are also contributory factors to the complex problem of corrosion.

A variety of corrosive substances occur in refineries (Table 3.1) and corrosion occurs in various forms in the refining process, such as pitting corrosion from acidic water, hydrogen embrittlement, and stress corrosion cracking from sulfide attack. Carbon steel is used for the

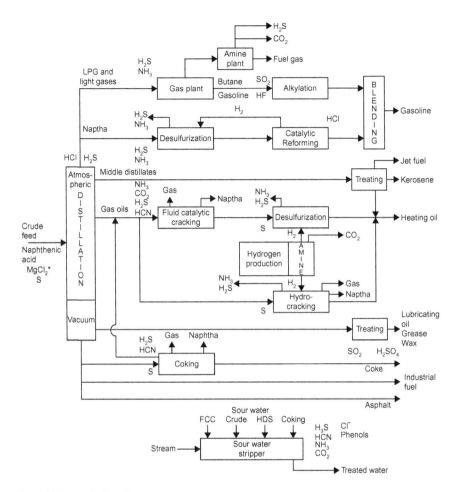

Figure 3.1 A typical refinery layout.

majority (>80%) of refinery components and is resistant to the most common forms of corrosion, particularly from hydrocarbon impurities at temperatures below 205°C (400°F). Other corrosive chemicals and environments prevent the use of carbon steel in all refinery units. Common replacement materials are low alloy steel containing chromium and molybdenum, with high-chromium stainless steel used for more corrosive environments. Nickel, titanium, and copper alloys are used in the most problematic areas where extremely high temperatures and/or very corrosive chemicals are present.

It is the purpose of this chapter to present the causes of corrosion that can occur in the various refinery units, and for more detail about

Table 3.1 Corrosive Substances that Occur in Refinery Feedstocks[a]	
Ammonia	Nitrogen in feedstocks combines with hydrogen to form ammonia (NH_3); ammonia is used for neutralization, which in turn may combine with other elements to form corrosive compounds, such as ammonium chloride (NH_4Cl).
Carbon	Not corrosive, but at high temperatures results in carburization that causes embrittlement or reduced corrosion resistance in some alloys.
Carbon dioxide	Occurs in steam reforming of hydrocarbon in hydrogen plants and to some extent in catalytic cracking; combines with moisture to form carbonic acid (H_2CO_3).
Chlorides	Present in the form of salts (such as magnesium chloride and calcium chloride) originating from crude oil, catalysts, and cooling water.
Cyanides	Usually generated in the cracking of high nitrogen feedstocks; when present, corrosion rates are likely to increase.
Hydrogen	Not typically a corrosive compound but can lead to blistering and embrittlement of steel; readily combines with other elements to produce corrosive compounds.
Hydrogen chloride	Formed through hydrolysis of magnesium chloride and calcium chloride, it is found in many overhead (vapor) streams; on condensation, it forms highly aggressive hydrochloric acid.
Naphthenic acids	A collective name for organic acids found in crude oils; structural aspects have not been fully characterized.
Oxygen	Originates in crude, aerated water, or packing gland leaks; aerial oxygen is used in furnace combustion and catalyst (FCC) regeneration units; results in high-temperature environments, which cause oxidation and scaling of metal surfaces of under-alloyed materials.
Phenols	Found primarily in sour water strippers but also in some crude oils and can contribute to acidic corrosion.
Polythionic acids	Sulfurous acids formed by the interaction of sulfides, moisture, and oxygen, and occurring when equipment is shut down.
Sulfur	Occurs in crude; causes high-temperature sulfidation of metals, and it combines with other elements to form aggressive compounds, such as various sulfides and sulfates and sulfurous, polythionic, and sulfuric acids.
Sulfuric acid	Used as a catalyst in alkylation plants and is formed in some process streams containing sulfur trioxide (SO_3) and water.

[a]The substances are listed alphabetically and not in any order of preference; the occurrence of such substance will be feedstock- and process-dependent.

these units, the reader is referred to earlier publications: Speight and Ozum (2002), Hsu and Robinson (2006), Gary et al. (2007), and Speight (2014).

3.2 STORAGE TANKS

Crude oil feedstock to the refinery and refinery products may be stored in aboveground steel storage tanks (*tank farm*), which act as a buffer between feedstock supply and demand for the refinery. The storage tanks can be as high as 50 and 100 ft or more wide, but tank size depends on the typical daily demand of crude oil, cycle time, safety stock, and working capacity: cycle time is the time between the production and delivery of a particular product.

The tanks are available in many shapes: vertical and horizontal cylindrical, open top and closed top, flat bottom, cone bottom, slope bottom, and dish bottom. Large tanks tend to be vertical cylindrical or to have rounded corners transitioning from vertical side wall to bottom profile, to safely withstand the hydrostatically induced pressure of contained liquid. Many tanks are covered with fixed or floating roofs to avoid evaporation and to minimize risk of fire. Aboveground storage tanks differ from underground storage tanks in the various regulations that are applied to such units. Most *container tanks* for handling liquids during transportation are designed to handle varying degrees of pressure. However, no matter what the tank shape, size, or function, corrosion can and will occur.

To combat corrosion, storage tanks include a liner on the internal floor and walls to prevent contact between the metal shell and any accumulated water or sediment. In fact, storage tanks suffer corrosion only if water is allowed to accumulate at the bottom with or without the sediment. In the United States, metal tanks in contact with soil and containing petroleum products must be protected from corrosion: the most effective and common corrosion control technique for steel in contact with soil is cathodic protection (Chapters 1 and 6).

3.3 DESALTING

As a first step in the refining process, to reduce corrosion, plugging, and fouling of equipment and to prevent poisoning the catalysts in processing units, contaminants must be removed by desalting (dehydration). Desalting is the means by which inorganic salts—that cause fouling or that hydrolyze and form corrosive acids—are largely removed. Often, chemicals are added in the form of demulsifiers to break the oil/water emulsion. The efficiency of the desalting operation is often directly related to the occurrence of corrosion in the distillation unit(s) as well as in other parts of the refinery.

3.3.1 Conventional Crude Oil

Produced crude is usually accompanied by hydrocarbon gases, hydrogen sulfide, carbon dioxide, and formation water. The variation of the solids content of individual crude oils solids originate in the formation from which the crude oil is produced and, as a result, very large day-to-day variations in the level of filterable solids are common. These solids can stabilize desalter emulsions, which generally contain high concentrations of the same solids that are found in raw crude samples (Kremer, 2000; Stark et al., 2002). In addition, the presence of scale—produced by corrosion and introduced during transportation—is also detrimental to the desalting operation.

Separators are used to degas produced crude and to remove the bulk of formation water. To meet the water content specified by pipeline companies, dehydrators are used to remove much of the remaining formation water and a portion of emulsified water. While salt content of produced crude depends primarily on salt content of formation water, salt content of dehydrated crude depends on the content of the basic sediment and water (BS&W: a technical specification of certain impurities in crude oil that originate from the reservoir formation; ASTM D4007, ASTM D7829). Conventional crude oil leaving the dehydrators typically contains 0.05−0.2% v/v BS&W, medium crude 0.1−0.4% v/v, and heavy crude 0.3−3% v/v. These values may be sufficiently low to meet the usual shipping specifications of pipeline operators, but dehydrators followed by field desalters are often necessary before the crude can be shipped to a refinery.

Thus, desalting reduces: (1) salt buildup and under-deposit corrosion in preheat exchangers, (2) corrosion in the flash zone, upper sections, and side-stream piping of the atmospheric column, and (3) corrosion in the upper sections of the atmospheric and vacuum columns. Desalting also decreases the amount of bottom sediment and water in the crude oil and the amount of suspended metal compounds going to downstream units via reduced crude (atmospheric resid) and vacuum resid (Gutzeit, 2008). Desalting has no effect on organic chlorides in crudes (Kronenberger, 1984; Barnett, 1988; Gutzeit, 2000; Lindemuth et al., 2001; Speight and Ozum, 2002; Choi, 2005; Mandal, 2005; Hsu and Robinson, 2006; Gary et al., 2007; Speight, 2014).

Inadequate desalting can cause fouling of heater tubes and heat exchangers throughout the refinery, which restricts product flow and heat transfer and leads to failures due to increased pressures and temperatures (dos Santos Liporace and de Oliveira, 2005). Corrosion, which occurs due to the presence of hydrogen sulfide, hydrogen chloride, naphthenic (organic) acids, and other contaminants in the crude oil, also causes equipment failure. Neutralized salts (ammonium chlorides and sulfides), when moistened by condensed water, can also cause corrosion. In addition, when elevated operating temperatures are used when desalting sour crudes, hydrogen sulfide will be present. There is the possibility of exposure to ammonia, dry chemical demulsifiers, caustics, and/or acids during this operation.

Chemicals used in the desalting operation improve overall desalting efficiency, reduce water and solids carryover with desalted crude, and reduce oil carry-under with brine effluent. Most desalting chemicals are demulsifiers that help break up the tight emulsion formed by the mix valve and produce relatively clean phases of desalted crude and brine effluent. If necessary, demulsifiers can be custom formulated for high water removal rates from crude oils, but at the cost of poor solids wetting and likely oil carry-under with the brine discharge. When formulated for high solids wetting rates, brine quality often decreases and water carryover with desalted crude increases.

Also, chemicals such as caustic soda are introduced into the desalting operation to neutralize acidic components, but uncontrolled use of caustic can have a detrimental effect. An excess of caustic can result in the formation of soap-type products due to the presence of fatty acids: the soap-like products stabilize the oil–water mixture and obstruct the separation process. Regular monitoring of the pH in the desalter water effluent allows for efficient dosing of caustic, which may result in a more efficient operation.

3.3.2 Heavy Crude Oil
Heavy crude oil (15–25° API) is being used more often as refinery feedstocks (Speight, 2011a,b, 2014). Inasmuch as there can be issues with conventional crudes in the desalter, anticipation of problems in the desalter from heavy feedstocks must be recognized. These oils often contain high levels of filterable solids, unstable asphaltene constituents (i.e., constituents likely to form a separate phase), and/or difficult-to-remove chloride salts. Since the filterable solids, asphaltene constituents, and salts are concentrated in the low-volatile (bottoms) fractions, the more difficult feedstocks tend to be: (1) heavy, conventionally produced crude oils, (2) extra heavy oil that has a low degree of mobility unless heated, and (3) tar sand bitumen diluted with lighter

hydrocarbons or synthetic crudes to meet pipeline gravity and viscosity specifications. However, there are also lower density crude oils (25–28° API) that can be problematic, owing to their high filterable solids contents, and only increased vigilance can overcome the impact of these feedstocks on desalter operations as well as on refinery operations in general (Waguespack and Healey, 1998; Speight, 2000, 2014; Stark and Asomaning, 2003).

Asphaltene instability causes desalter problems: the asphaltene constituents are known to stabilize water-in-oil emulsions, perhaps due to their concentration from the oil phase to the oil/water interface. The result can be BS&W carryover into the desalted crude oil and the appearance of asphaltene particulates in the desalter effluent water: both of which can cause sludge accumulation in crude storage tanks, as well as preheat fouling in heat exchanger fouling and pre-flash units, as well as foaming and corrosion in the atmospheric column.

Nonwater extractable chlorides that are not removed during desalting operations are typically defined as organic or inorganic chlorides that are not dissolved in emulsified water, removed by desalters, or indicated in normal, extractable chloride measurement methods. The identity of the chlorides varies with the specific feedstock (ASTM D4929), nevertheless, presence of these materials results in higher crude unit atmospheric column overhead system chloride loadings, increased overhead neutralizer demand, and higher overhead condensing, as well as a high potential for system corrosion (Table 3.2).

Maintaining desalter performance while processing the challenging heavy feedstocks requires careful attention to equipment design and

Table 3.2 Corrosion of Steel in Soil			
	Corrosion (mpy)	Type	Soil Resistivity (Ω/cm)
Typical soil	61	Moderately corrosive	1000–2000
Tidal marsh	100	Corrosive	500–1000
Clay	137	Very corrosive	Below 500
Sandy loam	21	Mildly corrosive	2000–10,000
Desert sand	5	Noncorrosive	Above 10,000
mpy, mils per year. Source: *Romanoff (1957).*			

operating conditions, which include: (1) wash water rate, (2) mixing energy, (3) temperature, and (4) mud-wash practice.

Increasing the *wash water rate* generally increases desalter performance in terms of the salt content and the BS&W of the desalted crude and desalter effluent water quality. It is often assumed that higher percentages of wash water result in wetter crude oil, but the opposite is true. When the wash water rate is increased, there are more water droplets that are closer together and the coalescing of small droplets into larger ones is facilitated. Increasing the wash water rate also dilutes the concentration of stabilizing molecules at the oil/water interface, which reduces the coalescing of droplets.

In terms of *mixing energy*, contaminant removal from crude oil requires the direct contact of the wash water droplets with the contaminant. If there is not enough mixing, then not all of the contaminants will be removed. If there is too much mixing, then very small water droplets are formed, producing an emulsion that cannot be fully resolved in the desalter vessel. Overmixing results in a high BS&W in the desalted crude oil and in poor salt removal, both of which lead to corrosion.

Temperature is a process variable that is not typically changed much in the desalter, but there are cases when favorable exchanger configurations can be used to increase desalter temperature. There are also cases in which changing the crude slate will change the heat balance on the tower and affect the desalter temperature. In general, raising the temperature will improve the oil/water separation in the desalter because the viscosity of the hydrocarbon decreases as the temperature increases. There are also naturally occurring materials at the oil/water interface that stabilize the desalter emulsion, but they are dissolved in the oil or water phase at higher temperatures, making the emulsion easier to break.

However, there are negative impacts to raising the desalter temperature: (1) asphaltene constituents can become unstable as the temperature is increased (Mitchell and Speight, 1973) and precipitated asphaltene constituents collect at the oil/water interface to stabilize the emulsion and cause oil under-carry and (2) water is more soluble in oil at higher temperatures, reducing the ability to dehydrate the crude oil and leading to high BS&W in the desalted crude. For various practical reasons, the upper limit of the desalter temperature is typically on the order of

155°C (310°F); most desalters run at temperatures substantially below the 155°C maximum, but the optimum operating temperature for the crude feedstock to be processed should be determined.

Mud-wash practice refers to the removal of contaminants such as sand, clay, iron sulfide, iron oxide, and other solids that settle to the bottom of the desalting vessel and form a *mud* or *sludge*. Heavy crude oil typically contains more filterable solids than conventional crude oil and tends to generate higher volumes of mud in the desalter. This reduces the working volume of the desalter, the water outlet can become partially blocked, and, in desalters with inlet headers in the bottom of the vessel, the inlet manifold can be partially plugged. The mud buildup will lead to increased oil in the effluent brine and poor desalter performance, leading to passage of undesirable materials through the desalter and corrosion. It is critical for desalters processing heavy feedstocks to incorporate a mud-wash system to remove mud from the bottom of the desalter. Furthermore, depending upon the properties of the heavy feedstock, it may be essential to perform the mud-wash operation at least once per day; if the mud is allowed to accumulate for several weeks, it becomes compacted and cannot be easily removed from the desalter.

3.3.3 Opportunity Crude Oil and Oil from Tight Shale

The characteristics of many heavy feedstocks (Speight, 2011a,b, 2014) include high solids levels, unstable asphaltene constituents, nonextractable chlorides, and considerable variability in one or more of these parameters for a given grade of crude oil. In particular, opportunity crudes typically have high levels of naphthenic acids, sulfur, and metals, and require more intensive processing to yield high-quality products (Kremer, 2006a,b). These crudes are medium to heavy (15–25° API) and vary considerably in properties, which relates to high (negative) impact on plant operations and equipment leading to elevated levels of downstream corrosion. Improved desalter operating procedures, chemical and monitoring programs, and constraint analysis can offer ways to mitigate the negative effects from such crudes (Gutzeit, 2008).

The challenges associated with the production of crude oil from tight shale deposits (sometimes erroneously called *shale oil*) are a function of their compositional complexities and the varied geological formations where they are found. These crude oils have a low molecular weight envelope but are waxy and reside in oil-wet formations. These properties create some of the main difficulties associated with refining, such as scale formation, salt deposition, paraffin wax deposits, destabilized asphaltene constituents, corrosion, and bacterial growth in storage.

In the case of oil from tight shale deposits, solids loading can be highly variable, leading to large shifts in solids removal performance. Sludge layers from the tank farm may cause severe upsets, including the growth of stable emulsion bands and intermittent increases of oil in the brine water. Agglomerated asphaltene constituents can enter the desalter from storage tanks or can flocculate in the desalter, leading to oil in the effluent brine.

Crude oil from tight shale deposits often contains high concentrations of hydrogen sulfide that require treatment with scavengers due to safety purposes. Amine-based scavengers often decompose as the crude oil is preheated through the hot preheat train and furnace, forming amine fragments. Ethanolamine (monoethanolamine), one of the most commonly used amines, readily forms an amine−chloride salt in the atmospheric tower. These salts deposit in the upper sections and, often, under-deposit corrosion is the major cause of failures in process systems because in the tower the under-salt corrosion rates can be 10−100 times faster than a general acidic attack. Mitigation strategies include: (1) controlling chloride to minimize the chloride occurrence in the tower overhead and (2) acidifying the desalter brine to increase removal of amines into the water phase.

3.4 DISTILLATION

After dewatering and desalting, the next step in the refining process is the separation of crude oil into various fractions or straight-run cuts by distillation in atmospheric and vacuum towers. The main fractions (*cuts*) obtained have specific boiling-point ranges and can be classified in order of decreasing volatility into gases, light distillates, middle distillates, gas oils, and residuum. The concentration of certain constituents by the distillation process can cause corrosion and sediment fouling, and can affect flow rates (Petkova et al., 2009). A properly designed distillation column can reduce the effects of these consequences, but corrosion and other effects are very prominent in reducing separation efficiency in the column.

Prior to entering the distillation unit, desalted crude feedstock is preheated and then flows to a direct-fired crude charge heater. There it is introduced into the distillation column at pressures slightly above atmospheric and at temperatures ranging from 345°C to 370°C (650–700°F), at which the lower boiling constituents flash into vapor. As the hot vapor rises in the tower, its temperature is reduced. Heavy fuel oil or residuum is taken from the bottom and, at successively higher points on the tower, the various major products are drawn off (Speight and Ozum, 2002; Hsu and Robinson, 2006; Gary et al., 2007; Speight, 2014).

However, despite a seemingly efficient desalting operation, corrosion agents can still appear in various streams during postdesalting (Blanco and Hopkinson, 1983; Piehl, 1988). For example, in the distillation unit, acid gases are formed, of which hydrogen sulfide is notorious. Steam, which is injected into the crude tower to improve the fractionation, condenses in the upper part of the unit. The hydrogen sulfide dissolves in the condensate and forms a weak acid that is known to cause stress corrosion cracking in the top section of the tower and in the overhead condenser. This may lead to frequent replacement of the tubing of the condenser and—in severe cases—to the replacement of the entire crude tower top.

In addition, the desalter wastewater (brine) and aqueous condensate from the overhead reflux drums of the main fractionating column contain appreciable levels of hydrogen sulfide, as does the hot-well water from the vacuum unit. The common practice is to strip hydrogen sulfide from the sour water (sour water stripping) before disposal. Fuel gas and the liquefied petroleum gas stream, which may contain hydrogen sulfide, are (should be) subjected to an amine wash to remove hydrogen sulfide before the caustic wash. The sour water, fuel gas, and rich amine containing hydrogen sulfide are potential sources for corrosion.

Typically, corrosion inhibitors and neutralizers, such as caustic soda or ammonia, are injected with the aim of increasing the pH of the sour water. The presence of various acidic gases and ammonia can result in solid salt depositing, from which ammonium bisulfide forms—this product is one of the main causes of alkaline sour water corrosion. Alkalinity (pH > 7.6) dramatically increases ammonium bisulfide corrosion and, as in desalting, the key to corrosion reduction is in accurate pH control. Proper neutralizer dosing will reduce not only corrosion but also chemical consumption. Reductions in the use of corrosion inhibitors of more than 15% have been reported.

Failure of condenser tubes constitutes the largest cause of outages in the distillation section, so the choice of tube material is accordingly crucial (Chapters 2 and 6). The thermal conductivity of the tubes has to be reasonably high, there must be sufficient ductility to expand into the tube plate, and the corrosion performance should be well understood. Three types of materials present themselves as being adequate to the task: (1) copper-base alloys, (2) stainless steels, and (3) titanium—each possesses its own merits and limitations (Chapter 2).

3.4.1 Hydrogen Chloride/Hydrogen Sulfide Corrosion

Corrosion in the atmospheric crude distillation unit overhead system stems primarily from the presence of hydrogen chloride. The most common source of hydrogen chloride (HCl, a low molecular weight volatile gas) is the decomposition of sodium chloride (NaCl), calcium chloride ($CaCl_2$), and magnesium chloride ($MgCl_2$) at temperatures exceeding $120°C$ ($250°F$), as well as from the decomposition of any organic chlorides. The hydrogen chloride moves into the crude unit overhead condensing systems where it is readily absorbed into condensing water.

Hydrogen chloride in the absence of water does not significantly corrode carbon steel (Sloley, 2013a,b). The overhead system of the crude tower condenses water, which absorbs the hydrogen chloride, creating hydrochloric acid. The water also absorbs ammonia (NH_3), which combines with hydrogen chloride and forms ammonium chloride (NH_4Cl). In situations in which the water returns to the vapor state, solid deposits of ammonium chloride form. This creates the potential for under-deposit corrosion, which occurs when corrosive salts form before a water phase is present. The strong acid hydrogen chloride reacts with ammonia and neutralizing amines to form salts that deposit on process surfaces. These salts are acidic and also readily absorb water from the vapor stream. The water acts as the electrolyte to enable these acid salts to corrode the surface and, typically, pitting occurs beneath these salts.

Various remedies are used to mitigate the acidic attack from condensed water containing hydrogen chloride, including neutralizing compounds like ammonia and organic amines, film-forming inhibitors, wash water systems, and careful control of temperature in the overhead circuit. To a lesser extent, chlorides can also enter the unit as entrained solids protected by an oil film (Chambers et al., 2011; Sloley, 2013a,b).

Generally, the segments of the distillation section that are susceptible to corrosion include (but may not be limited to) preheat exchanger (due to the presence of hydrogen chloride and hydrogen sulfide), preheat furnace and bottoms exchanger (hydrogen sulfide and sulfur compounds), atmospheric tower and vacuum furnace (hydrogen sulfide, sulfur compounds, and organic acids), vacuum tower (hydrogen sulfide and organic acids), and overhead (hydrogen chloride, hydrogen sulfide, and water). Where high sulfur (sour) crude oils are processed, severe corrosion can occur in furnace tubing and in both atmospheric and vacuum towers when metal temperatures exceed 235°C (450°F). Wet hydrogen sulfide will also cause cracks in steel. When processing high nitrogen crude oils, nitrogen oxides can form in the flue gases of furnaces and these gases are corrosive to steel when cooled to low temperatures in the presence of water.

There are three main ways to neutralize acidic aqueous solutions in the crude distillation unit: (1) by injecting a gaseous ammonia, (2) by injecting an ammonium hydroxide solution, and (3) by injecting neutralizing amine solutions (Jambo et al., 2002). Regardless of the neutralization technique applied, the pH is lower than the dew point of water. This adds more challenges in measuring pH when condensation occurs; this is the preferred region for the corrosion process to begin. Neutralization equations are:

$$HCl(aq) + NH_3(aq) \rightarrow NH_4Cl(aq)$$
$$HCl(aq) + RNH_2(aq) \rightarrow RNH_2 \bullet HCl(aq)$$

One concern with respect to neutralization is the difficulty of controlling the ammonia or amine flow rates, which depend on the varying levels of hydrogen chloride levels in the distillation unit. The neutralizer injection levels can be too low and the pH in the overhead can drop. Excess neutralizer levels, especially in the presence of hydrogen sulfide, contribute to precipitation of salts, such as ammonia or amine disulfides or chlorides. Once formed, these salts (molten or solid) deposit on pipe surfaces. Likewise, they can cause localized corrosion with a high rate of thickness loss. If salt formation occurs after condensation, then salt dissolution into water represents minimal corrosion.

3.5 NAPHTHENIC ACID CORROSION

In addition to acidic corrosion (by hydrogen chloride and hydrogen sulfide) at high temperatures, another form of acidic corrosion in the atmospheric and vacuum distillation units is a major concern (Chapter 1) (Kane and Cayard, 2002). In high acid (high-total acid number (TAN)) crude processing, acidity increases significantly in the overhead of atmospheric and vacuum units, as these crudes generally contain higher concentrations of salt, sediments, and sometimes organic chlorides. Desalting these crudes is difficult and, therefore, salt carryover to the overhead is more likely. Naphthenic acid corrosion and high-temperature crude corrosivity in general is a reliability issue in refinery distillation units. The presence of naphthenic acid and sulfur compounds considerably increases corrosion in the high-temperature parts of the distillation units.

In the vacuum column, preferential vaporization and condensation of naphthenic acids increase the acid content of the condensates. The naphthenic acids are most active at their boiling points, but the most severe corrosion generally occurs when the vapors condense to the liquid phase. In fact, the corrosion mechanism is mainly condensate corrosion and is directly related to the content, molecular weight, and boiling point of the naphthenic acid. Corrosion is typically severe at the condensing point, corresponding to high acid content and temperature.

Therefore, desalter management and corrosion control and monitoring are critical in the atmospheric distillation unit and the vacuum distillation unit, as well as in the visbreaking unit (Gutzeit, 2008). In fact, residence time of the crude in the desalter, the wash-water injection rate, the mixing valve differential pressure, and efficacy of the emulsion breaker all play vital roles in corrosion management.

3.6 COKING

Coking is a severe method of thermal cracking used to upgrade heavy residuals into lighter products or distillates. Coking produces straight-run gasoline (coker naphtha) and various middle-distillate fractions used as catalytic cracking feedstock and a residue (coke). The two most common processes are delayed coking and continuous (contact or fluid) coking. Three typical types of coke obtained are: (1) sponge coke, (2) honeycomb coke, and (3) needle coke, depending upon the reaction mechanism, time, temperature, and crude feedstock. On the other hand, visbreaking, which is often referred to as a mild thermal cracking process, insofar as the reactions are not allowed to completion is also a high-temperature process.

In these processes, especially during processing of high-sulfur crudes, corrosion can occur when metal temperatures are between 235°C and 485°C (450°F and 900°F). Above 485°C (900°F), coke forms a protective layer on the metal, but the furnace, soaking drums, lower part of the tower, and high-temperature exchangers are usually subject to corrosion. Hydrogen sulfide corrosion in coking and visbreaking can also occur when temperatures are not properly controlled above 485°C (900°F).

Continuous thermal changes can lead to the bulging and cracking of reactor shells. For example, in the coking process, temperature control must often be held within a 5−10°C (10−20°F) range, as high temperatures will produce coke that is too hard to remove from the reactor surface or coke drum. Conversely, temperatures that are too low will result in a slurry in the reactor (especially in the visbreaker, where the liquid phase is the dominant phase) with a high content of asphaltene constituents.

3.7 CATALYTIC CRACKING

Catalytic cracking is similar to thermal cracking except that a catalyst (a material that assists the chemical reaction but does not take part in it) is employed to facilitate the conversion of the higher molecular weight feedstock constituents into lower molecular weight products. Use of a catalyst in the cracking reaction increases the yield of improved-quality products under much less severe operating conditions than in thermal cracking. The typical catalytic cracking unit is a fluidized bed system; other units include the *moving-bed catalytic cracking process*. In the *Thermofor catalytic cracking unit*, the preheated feedstock flows by gravity through the catalytic reactor bed.

In the fluid catalytic cracking unit, hydrogen sulfide is evolved from sulfur impurities from the feedstock and concentrates in the produced hydrocarbon gas. The reactor outlet products, which are sent to the main fractionator, can also contain hydrogen sulfide (approximately 0.4% v/v).

The gas from the sponge absorber contains the majority of the hydrogen sulfide (9.5% v/v) and all of the carbon dioxide (CO_2) entrained in the regenerated catalyst as inert. These two acid gases are removed from the absorber gas via amine treatment before being sent to the refinery fuel–gas pool.

The sour water (a potential corrosion risk) from the fractionator overhead receiver is sent to the wet-gas compressor first-stage discharge as wash water. Fresh makeup wash water is added to the sour water as required. Corrosion-control chemicals are injected into the wash water and provide corrosion protection in the circuit. The aqueous condensate in the high-pressure separator contains hydrogen sulfide. Sour water from the high-pressure separator is sent to the sour water stripper to remove hydrogen sulfide.

Regular sampling and testing of the feedstock, product, and recycle streams should be performed to assure that the cracking process is proceeding as intended and that no contaminants have entered the process stream. Corrosive materials or deposits in the feedstock can foul gas compressors and inspections of critical equipment, including pumps, compressors, furnaces, and heat exchangers, which should be conducted as needed. When processing sour crude, corrosion may be expected when temperatures are below 485°C (900°F). Corrosion takes place where both liquid and vapor phases exist and in areas subject to local cooling, such as nozzles and platform supports.

When processing high nitrogen feedstocks, the formation of ammonia and cyanide may occur, subjecting carbon steel equipment in the fluid catalytic cracker overhead system to corrosion, cracking, or hydrogen blistering. These effects may be minimized by water wash or corrosion inhibitors; water wash may also be used to protect overhead condensers in the main column subjected to fouling from ammonium hydrosulfide (NH_4HS). Inspections should include checking for leaks due to erosion or other malfunctions, such as catalyst buildup on the expanders, coking in the overhead feeder lines from feedstock residues, and other unusual operating conditions.

3.8 HYDROPROCESSES

In hydrotreating operations, many of the processes require hydrogen generation to provide for a continuous supply. Because of the operating temperatures and the presence of hydrogen, the hydrogen sulfide content of the feedstock must be strictly controlled to a minimum to reduce corrosion. Hydrogen chloride may form and condense as hydrochloric acid in the lower-temperature parts of the unit and ammonium hydrosulfide may form in high-temperature, high-pressure units.

In hydrocracking, because of the operating temperatures and the presence of hydrogen, the hydrogen sulfide content of the feedstock must be strictly controlled to a minimum to reduce the possibility of severe corrosion. Corrosion by carbon dioxide in the presence of water in areas of condensation must also be considered because of the formation of carbonic acid:

$$CO_2 + H_2O \rightarrow H_2CO_3$$

Hydrogen sulfide and ammonia are formed by the decomposition of organic sulfur and nitrogen impurities in the feedstock. These two reaction products combine to form ammonium salts, which can solidify and precipitate as the reactor effluent is cooled. Likewise, ammonium chloride may be formed if any chloride is in the system. A water wash to reactor effluent dissolves these salts before they can precipitate. Sour water containing high levels of dissolved hydrogen sulfide should not be released to the atmosphere. Off-gas and recycled gas containing high levels of hydrogen sulfide should also be treated in an amine unit.

During the last several decades, ebullated-bed residue hydrocracking has gained interest due to its capability to produce high-quality, low-boiling and mid-boiling distillates from heavy feedstocks in an economically effective way. Major economical drivers for ebullated-bed hydrocracker processes are run length, maintenance costs, and, most importantly, the achieved conversion. However, feedstock conversion rates, leading to fouling and the subsequent corrosion by deposition of asphaltene constituents or reacted asphaltene constituents, increase more rapidly with rising temperatures as compared with the hydrogen-saturation reactions that inhibit sediment formation. Accordingly, increasing temperature and conversion rates above certain limits and beyond the optimal operational window will lead to uncontrolled sedimentation and coke generation. These generated foulants will be deposited in critical plant sections leading to corrosion or will cause problems related to sediment specification for heavy fuel oil (Kunnas et al., 2010).

In both hydrotreating and hydrocracking processes, hydrogen damage (hydrogen-assisted cracking is the generic name given to a large number of metal degradation processes due to interaction with hydrogen) can occur in carbon steel through the diffusion of atomic hydrogen into the metal, where it combines with the carbon in the iron carbide form (Fe_3C or $Fe_2C = Fe$) to form methane and to eliminate the pearlite constituent. Hydrogen damage is a potential problem in reactor pressure vessels in hydrogen service: the concern is that such subcritical cracks do not reach a critical size for failure. The important aspect of this type of corrosion is the ability to detect and accurately measure the depth of such cracks (beneath stainless steel cladding) so that accurate predictions of reactor life (or reactor failure) can be made.

3.9 PRODUCT IMPROVEMENT PROCESSES

The term *product improvement processes* includes the various processes that convert the raw product into a saleable product that meets the various specifications required by the buyer. These processes change the properties of the product relative to the feedstock, and such processes are conducive to expansion of the utility of petroleum products and to sales.

3.9.1 Solvent Processes

Solvent deasphalting processes are a major part of refinery operations (Speight and Ozum, 2002; Hsu and Robinson, 2006; Gary et al., 2007; Speight, 2014) and are not often appreciated for the tasks for which they are used. In the solvent deasphalting processes, an alkane is injected into the feedstock to cause the high molecular weight constituents and polar constituents to precipitate. Propane (or sometimes a propane/butane mixture) is extensively used for deasphalting and produces a deasphalted oil and propane deasphalter asphalt (PD tar) (Speight and Ozum, 2002; Hsu and Robinson, 2006; Gary et al., 2007; Speight, 2014). Propane has unique solvent properties; at lower temperatures (38−60°C; 100−140°C) paraffins are very soluble in propane and at higher temperatures (approximately 93°C; 200°F) all hydrocarbons are almost insoluble in propane.

Solvent dewaxing is used to remove wax from either distillate or residual base stocks at any stage in the refining process. There are several processes in use for solvent dewaxing, but all have the same general steps, which are: (1) mixing the feedstock with a solvent, (2) precipitating the wax from the mixture by chilling, and (3) recovering the solvent from the wax and dewaxed oil for recycling by distillation and steam stripping. Two solvents are usually used: toluene, which dissolves the oil and maintains fluidity at low temperatures, and methyl ethyl ketone, which dissolves little wax at low temperatures and acts as a wax precipitating agent. Other solvents that are sometimes used include benzene, methyl isobutyl ketone, propane, petroleum naphtha, ethylene dichloride, methylene chloride, and sulfur dioxide. In addition, there is a catalytic process used as an alternative to solvent dewaxing.

To prevent wax from depositing on the walls of the inner pipe, blades or scrapers extending the length of the pipe and fastened to a central rotating shaft scrape off the wax. Slow chilling reduces the temperature of the waxy oil solution to 2°C (35°F), and then faster chilling reduces the temperature to the approximate pour point required in the dewaxed oil. The waxy mixture is pumped to a filter case into which the bottom half of the drum of a rotary vacuum filter dips. The drum (8 ft in diameter, 14 ft long), covered with filter cloth, rotates continuously in the filter case. A vacuum within the drum sucks the solvent and the oil dissolved in the solvent through the filter cloth and into the drum. Wax crystals collect on the outside of the drum to form a wax cake and, as the drum rotates, the cake is brought above the surface of the liquid in the filter case and under sprays of ketone that wash oil out of the cake and into the drum. A knife-edge scrapes off the wax, and the cake falls onto the conveyor and is moved away from the filter by the rotating scroll.

Solvent extraction is used to prevent corrosion, protect the catalyst in subsequent processes, and improve finished products by removing unsaturated, aromatic hydrocarbons from lubricant and grease stocks. The solvent extraction process separates aromatics, naphthenes, and impurities from the product stream by dissolving or precipitating them. The feedstock is first dried and then treated using a continuous countercurrent solvent treatment operation. In one type of process, the feedstock is washed with a liquid in which the substances to be removed are more soluble than in the desired resultant product. In another process, selected solvents are added to cause impurities to precipitate out of the product. In the adsorption process, highly porous solid materials collect liquid molecules on their surfaces.

The solvent is separated from the product stream by heat, evaporation, or fractionation, and residual trace amounts are subsequently removed from the raffinate by steam stripping or vacuum flashing. Electric precipitation may be used for separation of inorganic compounds. The solvent is then regenerated to be used again in the process. The most widely used extraction solvents are phenol, furfural, and cresylic acid. Other solvents less frequently used are liquid sulfur dioxide, nitrobenzene, and 2,2′-dichloroethyl ether.

3.9.2 Reforming Processes

Thermal reforming is less effective and less economical than catalytic processes are, and it has largely been supplanted. As it used to be practiced, a single-pass operation was employed at temperatures in the range of 540–760°C (1000–1140°F) and pressures of about 500–1000 psi (34–68 atmospheres). The degree of octane number improvement depended on the extent of conversion but was not directly proportional to the extent of crack per pass. However, at very high conversions, the production of coke and gas became prohibitively high. The gases produced were generally olefins and the process required either a separate gas polymerization operation or one in which C3 to C4 gases were added back to the reforming system.

In the catalytic reformer, operating procedures should be developed to ensure control of hot spots during start-up. Safe catalyst handling is very important. Care must be taken not to break or crush the catalyst when loading the beds, as the small fines will plug up the reformer screens. Precautions against dust when regenerating or replacing the catalyst should also be considered. Also, water wash should be considered where stabilizer fouling has occurred due to the formation of ammonium chloride and iron salts. Ammonium chloride may form in pretreater exchangers and cause corrosion and fouling. Hydrogen chloride from the hydrogenation of chlorine compounds may form acid or ammonium chloride salt.

In the steam reforming process, the potential exists for burns from hot gases and superheated steam should a release occur. Inspections and testing are necessary where the possibility exists for valve failure due to contaminants in the hydrogen. Carryover from caustic scrubbers should be controlled to prevent corrosion in preheaters. Chlorides from the feedstock or steam system should be prevented from entering reformer tubes and contaminating the catalyst.

Feed-water supply is an important part of steam generation and the water must be free of contaminants, including minerals and dissolved impurities that can damage the system or affect its operation. Suspended materials such as silt, sewage, and oil, which form scale and sludge, must be coagulated or filtered out of the water. Dissolved gases, particularly carbon dioxide and oxygen, cause boiler corrosion and are removed by deaeration and treatment. Dissolved minerals—including metallic salts and carbonates that cause scale, corrosion, and turbine blade deposits—are treated with lime or soda ash to precipitate them from the water. Recirculated cooling water must also be treated for hydrocarbons and other contaminants.

Recirculated cooling water must be treated to remove impurities and dissolved hydrocarbons. Typically, the water is saturated with oxygen from being cooled with air and the chances for corrosion are increased. One means of corrosion prevention is the addition of a material to the cooling water that forms a protective film on pipes and other metal surfaces.

3.9.3 Alkylation Processes

The combination of olefins with paraffins to form higher isoparaffins is termed alkylation. Sulfuric acid, hydrogen fluoride, and aluminum chloride are the general catalysts used commercially. Sulfuric acid is used with propylene and higher boiling feeds, but not with ethylene, because it reacts to form ethyl hydrogen sulfate. The acid is pumped through the reactor and forms an air emulsion with reactants, and the emulsion is maintained at 50% acid. The rate of deactivation varies with the feed and isobutane charge rate. Butene feeds cause less acid consumption than the propylene feeds.

In cascade type sulfuric acid (H_2SO_4) alkylation units, the feedstock (propene, butene, pentene, and fresh isobutane) enters the reactor and contacts the concentrated sulfuric acid catalyst (in concentrations of 85−95% for good operation and to minimize corrosion). The reactor is divided into zones, with olefins fed through distributors to each zone, and the sulfuric acid and isobutane flowing over baffles from zone to zone.

Some corrosion and fouling in sulfuric acid units may occur from the breakdown of sulfuric acid esters or where caustic is added for neutralization. These esters can be removed by treating the mixture with fresh acid and washing it with hot water. To prevent corrosion from hydrofluoric acid, the acid concentration inside the process unit should be maintained above 65% and the moisture should be below 4%.

3.9.4 Polymerization Processes

Polymerization processes are processes by which olefin gases are converted into liquid products that may be suitable for gasoline (polymer gasoline) or other liquid fuels. The feedstock usually consists of propene and butene isomers from cracking processes or may even be selective olefins for dimer, trimer, or tetramer production.

Phosphates are the principal catalysts used in polymerization units; the commercially used catalysts are liquid phosphoric acid, phosphoric acid on kieselguhr, copper pyrophosphate pellets, and phosphoric acid film on quartz. The latter is the least active, but the most used and the easiest to regenerate simply by washing and recoating; the serious disadvantage is that tar must occasionally be burned off the support. The process using a liquid phosphoric acid catalyst is far more responsive to attempts to raise production by increasing temperature than the other processes.

The potential for an uncontrolled exothermic reaction exists should loss of cooling water occur. Severe corrosion leading to equipment failure will occur should water make contact with the phosphoric acid, such as during water washing at shutdowns. Corrosion may also occur in piping manifolds, reboilers, exchangers, and other locations where acid may settle out.

The austenitic stainless steels of chromium−nickel−molybdenum (and to a lesser extent, chromium−nickel) compositions are commonly used for handling phosphoric acid solutions within the limits of concentration, temperature, aeration, and purity for which they are suitable.

3.9.5 Isomerization Processes

Aluminum chloride was the first catalyst used to isomerize butane, pentane, and hexane. Since then, supported metal catalysts have been developed for use in high-temperature processes that operate in the range of 370–480°C (700–900°F) and 300–750 psi, while aluminum chloride plus hydrogen chloride are universally used for the low-temperature processes. A nonregenerable aluminum chloride catalyst is employed with various carriers in a fixed-bed or liquid contactor. Platinum or other metal catalyst processes utilize fixed-bed operations and can be regenerable or nonregenerable. The reaction conditions vary widely depending on the particular process and the feedstock: process temperatures vary from 40°C to 480°C (100°F to 900°F) and pressures from 150 to 1000 psi.

Isomerization processes are to provide additional feedstock for alkylation units or high octane fractions for gasoline blending. Straight-chain paraffins (n-butane, n-pentane, n-hexane) are converted to their respective isocompounds by continuous catalytic (aluminum chloride, noble metals) processes. In the isomerization process, if the feedstock is not completely dried and desulfurized, the potential exists for acid formation, leading to catalyst poisoning and metal corrosion. Water or steam must not be allowed to enter areas where hydrogen chloride is present. Precautions are needed to prevent hydrogen chloride from entering sewers and drains.

3.9.6 Caustic Treating Processes

The caustic tower is a vertical gas/liquid contactor that is pressurized and operated between 30°C and 50°C (86°F and 122°F). The typical caustic tower (Figure 3.2) has three to four stages, starting with the top (water-wash) stage, the second (strong-caustic) stage, and then the bottom (intermediate- and weak-caustic) stages. Each stage has a liquid reservoir at the bottom. Gas/liquid contacting is enhanced by recirculating the caustic from the reservoir to the top of that stage. Part of the reservoir is cascaded down to the next stage. In the bottom stage, most of the free caustic has been consumed, and the weak caustic is loaded with sulfides, carbonates, and hydrocarbons. A portion of the weak liquor is recirculated in the bottom stage, and the remainder is discharged as spent caustic.

A layer of hydrocarbon oil may float on top of each caustic reservoir. The caustic tower should be designed to avoid retaining this red oil since higher residence time increases polymerization. The intermediate section sumps may be designed with a standpipe for down flow to the next section. The standpipe allows the oil floating on the caustic to

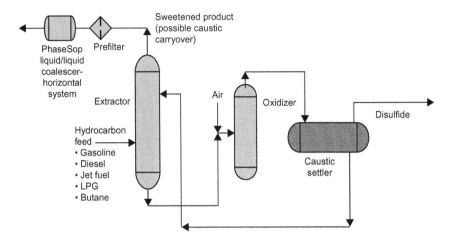

Figure 3.2 Schematic of a caustic treating process.

exit to the next stage, thus minimizing residence time. The bottom section sump is designed to allow skimming of the oil by the operators. Oil removed on a routine and frequent basis is typically easy to separate and less prone to cause fouling and corrosion than aged oil. The polymerization reaction is temperature dependent, so the warmer the tower, the faster the red oil formation rate. This is another reason that excessive reheating of the feed should be avoided.

Some towers do not have reservoirs with circulation pumps between the stages. Without the reservoirs, the red oil cannot be removed from within the tower. This can lead to a heavier hydrocarbon load on the tower and in the spent caustic. After the lowest stage, the corrosive spent caustic collects in the bottom of the tower. In some towers, a separate tap in the bottom is used to remove the floating red oil layer, so that it does not continue to polymerize. Others may drain the bottom completely on a periodic basis. Most towers also send the mixed spent caustic to a gravity-separation drum to skim the oil.

Caustic towers at ethylene plants remove acid gases, hydrogen sulfide, and carbon dioxide from ethylene gas. The spent caustic from these towers contains mercaptans and sulfides, which are reactive, odorous, and corrosive. Accordingly, the spent caustic requires special handling and treatment before being discharged to a conventional wastewater treatment plant. Often, the spent caustic is most commonly treated in an oxidation reactor (wet air oxidation) that converts the corrosive sulfides into oxidation products such as sulfate ions. Other components and contaminants present in the spent caustic can affect the wet air oxidation systems.

The discharged spent caustic passes to a separation drum where caustic, entrained gases, and entrained oils are separated by gravity. The spent caustic then passes through a depressurization valve and into a degassing drum where the evolved gases from depressurization are removed. The spent caustic is then routed to a gasoline washing step or to a storage tank. The "gasoline" wash is usually pygas (from the quench water tower), steam cracked naphtha (from the primary fractionator), or some other stream or combination.

The discharged spent caustic passes to a separation drum where caustic, entrained gases, and entrained oils are separated by gravity. The spent caustic then passes through a depressurization valve and into a degassing drum where the evolved gases from depressurization are removed. The spent caustic is then routed to a gasoline washing step or to a storage tank. Residence time in the storage tank is typically 24 h or longer. This tank serves as an equalization tank and is the feed tank for the wet air oxidation system. The configuration and operation of this tank can have an impact on the wet air oxidation system operations.

Over several years of operation, a heavy organic sludge that is formed from the aged red oil will build in the bottom of a nonagitated tank and can have a depth of up to 3 ft or more. As this depth increases, slugs of the oily sludge might be drawn into the wet air oxidation feed tap. The location of tank inlet ports can also have an effect, as the risk of drawing an oil slug increases if an inlet stream disturbs the bottom oil sludge. Typically, the spent caustic draw line is located a short distance from the bottom of the tank. Periodically cleaning the tank to remove the sludge buildup will reduce the risk of heavy oil slugs entering the wet air oxidation system.

REFERENCES

ASTM D4007, 2012. e1 Standard test method for water and sediment in crude oil by the centrifuge method (laboratory procedure). Annual Book of Standards. ASTM International, West Conshohocken, PA.

ASTM D4929, 2012. Standard test methods for determination of organic chloride content in crude oil. Annual Book of Standards. ASTM International, West Conshohocken, PA.

ASTM D7829, 2012. Standard guide for sediment and water determination in crude oil. Annual Book of Standards. ASTM International, West Conshohocken, PA.

Ayello, F., Robbins, W., Richter, S., Nešić, S., 2011. Crude oil chemistry effects on inhibition of corrosion and phase wetting. Paper No. 11060. In: Proceedings of the Corrosion 2011. NACE International, Houston, TX.

Barnett, J.W., 1988. Desalters can remove more than salts and sediment. Oil Gas J. 86 (4), 15.

Blanco, F., Hopkinson, B., 1983. Experience with naphthenic acid corrosion in refinery distillation process units. Paper No. 99. In: Proceedings of the Corrosion 83. NACE International, Houston, TX.

Chambers, B., Srinivasan, S., Yap, K.M., Yunovich, M., 2011. Corrosion in crude distillation unit overhead operations: a comprehensive review. In: Proceedings of the Corrosion 2011. NACE International. 13–17 March, Houston, TX.

Choi, D.W., 2005. How to buy and operate desalters. Hydrocarb. Process. 79 (3), 75–78.

Dettman, H.D., Li, N., Wickramasinghe, D., Luo, J., 2010. The influence of naphthenic acid and sulfur compound structure on global crude corrosivity under vacuum distillation conditions. In: Proceedings. COQA/CCQTA Joint Meeting. 10–11 February, New Orleans, LO.

dos Santos Liporace, F., de Oliveira, S.G., 2005. Real time fouling diagnosis and heat exchanger performance. In: Müller-Steinhagen, H., Malayeri, M.R., Watkinson, A.P. (Eds.), Proceedings of the 6th International Conference on Heat Exchanger Fouling and Cleaning: Challenges and Opportunities. ECI Symposium Series, vol. RP2. 5–10 June, Engineering Conferences International, Kloster Irsee, Germany.

Gary, J.G., Handwerk, G.E., Kaiser, M.J., 2007. Petroleum Refining: Technology and Economics, fifth ed. CRC Press, Taylor & Francis Group, Boca Raton, FL.

Gutzeit, J., 2000. Effect of organic chloride contamination of crude oil on refinery corrosion. In: Proceedings of the Corrosion 2000. NACE International, Houston, TX.

Gutzeit, J., 2008. Controlling crude unit overhead corrosion by improved desalting. Hydrocarb. Process. 82 (2), 119.

Hsu, C.S., Robinson, P.R. (Eds.), 2006. Practical Advances in Petroleum Processing, vols. 1 and 2. Springer Science, New York, NY.

Jambo, H.C.M., Freitas, D.S., Ponciano, J.A.C., 2002. Ammonium hydroxide injection for overhead corrosion control in a crude distillation unit. In: Proceedings of the International Corrosion Congress, Granada, Spain, September.

Kane, R.D., 2006. Corrosion in petroleum refining and petrochemical operations. Corrosion: environments and industries. ASM Handbook, vol. 13C. ASM International, Materials Park, OH, pp. 967–1014.

Kane, R.D., Cayard, M.S., 2002. A comprehensive study on naphthenic acid corrosion. In: Proceedings of the Corrosion 2002. NACE International, Houston, TX.

Kremer, L.N. 2000. Challenges to desalting heavy crude oil. In: Proceedings of the International Conference on Refinery Processing, 2000 AIChE Spring National Meeting. 5–9 March, Atlanta, GA.

Kremer, L.N., 2006a. Crude oil management: reduce operating problems while processing opportunity crudes. In: Proceedings of the International Conference on Refinery Processing, AIChE Spring National Meeting. 23–27 April, Orlando, FL.

Kremer, L.N. 2006b. Controlling quality variations in the feed to desalters. In: Proceedings of the International Conference on Refinery Processing, 2006 AIChE Spring National Meeting. 23–27 April, Orlando, FL.

Kronenberger, D.L., 1984. Corrosion problems associated with the desalting difficulties of maya and other heavy crudes. Paper No. 128. In: Proceedings of the Corrosion 84. NACE International, Houston, TX.

Kunnas, J., Ovaskainen, O., Respini, M., 2010. Mitigate fouling in ebullated-bed hydrocrackers. Hydrocarb. Process. 10 (1).

Lindemuth, P.M., Lessard, R.B., Lozynski, M., 2001. Improve desalter operations. Hydrocarb. Process. 75 (9), 67.

Mandal, K.K., 2005. Improve desalter control. Hydrocarb. Process. 79 (4), 77–82.

Mitchell, D.L., Speight, J.G., 1973. The solubility of asphaltenes in hydrocarbon solvents. Fuel 52, 149.

Petkova, N., Angelova, M., Petkov, P., 2009. Establishing the reasons and type of the enhanced corrosion in the crude oil atmospheric distillation unit. Petrol. Coal 51 (4), 286–292.

Piehl, R.L., 1988. Naphthenic acid corrosion in crude distillation units. Mater. Perform. 27 (1), 37–43.

Romanoff, M., 1957. Underground Corrosion, Circular No. 579, National Bureau of Standards, Washington, DC.

Sloley, A.W., 2013a. Mitigate fouling in crude unit overhead part 1. Hydrocarb. Process. 92 (9), 73–81.

Sloley, A.W., 2013b. Mitigate fouling in crude unit overhead part 2. Hydrocarb. Process. 92 (11), 73–75.

Speight, J.G., 2000. The Desulfurization of Heavy Oils and Residua, second ed. Marcel Dekker Inc., New York, NY.

Speight, J.G., 2011a. An Introduction to Petroleum Technology, Economics, and Politics. Scrivener Publishing, Salem, MA.

Speight, J.G., 2011b. The Refinery of the Future. Gulf Professional Publishing, Elsevier, Oxford, UK.

Speight, J.G., 2014. The Chemistry and Technology of Petroleum, fifth ed. CRC Press, Taylor & Francis Group, Boca Raton, FL.

Speight, J.G., Ozum, B., 2002. Petroleum Refining Processes. Marcel Dekker Inc., New York, NY.

Stark, J.L., Asomaning, S., 2003. Crude oil blending effects on asphaltene stability in refinery fouling. Pet. Sci. Technol. 21 (3 & 4), 569–579.

Stark, J.L., Nguyen, J., Kremer, L.N., 2002. Crude stability as related to desalter upsets. In: Proceedings of the Fifth International Conference on Refinery Processing, 2002 AIChE Spring National Meeting. 11–14 March, New Orleans, LO.

Tebbal, S., 1999. Critical review of naphthenic acid corrosion. Paper No. 380. In: Proceedings of the Corrosion 99. NACE International, Houston, TX.

Tebbal, S., Kane, R.D., 1996. Review of critical factors affecting crude corrosivity. Paper No. 607. In: Proceedings of the Corrosion 96. NACE International, Houston, TX.

Tebbal, S., Kane, R.D., 1998. Assessment of crude oil corrosivity. Paper No. 578. In: Proceedings of the Corrosion 98. NACE International, Houston, TX.

Tebbal, S., Kane, R.D., Yamada, K., 1997. Assessment of the corrosivity of crude fractions from varying feedstocks. Paper No. 498. In: Proceedings of the Corrosion 97. NACE International, Houston, TX.

Waguespack, K.G., Healey, J.F., 1998. Manage crude oil quality for refining profitability. Hydrocarb. Process. 77 (9), 133–140.

FURTHER READING

Ajeel, S.A., Abdullatef Ahmed, M.A., 2008. Study of the synergy effect on erosion–corrosion in oil pipes. Eng. Tech. 26 (9), 1068.

Barbouteau, L., Dalaud, R., 1972. In: Nonhebel, G. (Ed.), Gas Purification Processes for Air Pollution Control. Butterworth and Co., London, UK (Chapter 7).

Bartoo, R.K., 1985. In: Newman, S.A. (Ed.), Acid and Sour Gas Treating Processes. Gulf Publishing, Houston, TX.

Curry, R.N., 1981. Fundamentals of Natural Gas Conditioning. PennWell Publishing Co., Tulsa, OK.

Fulker, R.D., 1972. In: Nonhebel, G. (Ed.), Gas Purification Processes for Air Pollution Control. Butterworth and Co., London, UK (Chapter 9).

Jou, F.Y., Otto, F.D., Mather, A.E., 1985. In: Newman, S.A. (Ed.), Acid and Sour Gas Treating Processes. Gulf Publishing Company, Houston, TX (Chapter 10).

Katz, D.K., 1959. Handbook of Natural Gas Engineering. McGraw-Hill Book Company, New York, NY.

Kohl, A.L., Nielsen, R.B., 1997. Gas Purification. Gulf Publishing Company, Houston, TX.

Kohl, A.L., Riesenfeld, F.C., 1985. Gas Purification, fourth ed. Gulf Publishing Company, Houston, TX.

Lachet, V., Béhar, E., 2000. Industrial perspective on natural gas hydrates. Oil Gas Sci. Technol. 55, 611–616.

Maddox, R.N., 1974. Gas and Liquid Sweetening, second ed. Campbell Publishing Co., Norman, OK.

Maddox, R.N., 1982. Gas conditioning and processing, vol. 4. Gas and Liquid Sweetening. Campbell Publishing Co., Norman, OK.

Maddox, R.N., Bhairi, A., Mains, G.J., Shariat, A., 1985. In: Newman, S.A. (Ed.), Acid and Sour Gas Treating Processes. Gulf Publishing Company, Houston, TX (Chapter 8).

Mody, V., Jakhete, R., 1988. Dust Control Handbook. Noyes Data Corp., Park Ridge, NJ.

Mokhatab, S., Poe, W.A., Speight, J.G., 2006. Handbook of Natural Gas Transmission and Processing. Elsevier, Amsterdam, The Netherlands.

Newman, S.A., 1985. Acid and Sour Gas Treating Processes. Gulf Publishing, Houston, TX.

Nonhebel, G., 1964. Gas Purification Processes. George Newnes Ltd., London, UK.

Pitsinigos, V.D., Lygeros, A.I., 1989. Predicting $H_2S–MEA$ equilibria. Hydrocarb. Process. 58 (4), 43–44.

Polasek, J., Bullin, J., 1985. In: Newman, S.A. (Ed.), Acid and Sour Gas Treating Processes. Gulf Publishing Company, Houston, TX (Chapter 7).

Rooney, C.P., Dupart, M.S., 2000. Corrosion in alkanolamine plants: causes and minimization. In: Proceedings of the Corrosion 2000. NACE International, Orlando, FL.

Soud, H., Takeshita, M., 1994. FGD Handbook. No. IEACR/65. International Energy Agency Coal Research, London, UK.

Speight, J.G., 2007. Natural Gas: A Basic Handbook. GPC Books, Gulf Publishing Company, Houston, TX.

Speight, J.G., 2013. The Chemistry and Technology of Coal, third ed. CRC Press, Taylor & Francis Group, Boca Raton, FL.

Staton, J.S., Rousseau, R.W., Ferrell, J.K., 1985. Regeneration of physical solvents in conditioning gases from coal. Chapter 5. In: Newman, S.A. (Ed.), Acid and Sour Gas Treating Processes. Gulf Publishing Company, Houston, TX (Chapter 5).

Ward, E.R., 1972. In: Nonhebel, G. (Ed.), Gas Purification Processes for Air Pollution Control. Butterworth and Co., London, UK (Chapter 8).

Zhang, L.Q., Shi, L.B., Zhou, Y., 2007. Formation prediction and prevention technology of natural gas hydrate. Nat. Gas Technol. 1 (6), 67–69.

CHAPTER 4

Corrosion in Gas Processing Plants

4.1 INTRODUCTION

Gas processing (gas treating, gas refining) (Mokhatab et al., 2006; Speight, 2007, 2014) consists of separating all of the various hydrocarbons and fluids from the pure natural gas or refinery gas. Because of, among other issues, the potential for pipeline corrosion, major transportation pipelines impose restrictions on the makeup of the natural gas that is allowed into the pipeline. That means that before the natural gas can be transported, it must be purified and corrosive constituents must be removed. Although the processing of natural gas is in many respects less complicated than the processing and refining of crude oil, it is equally as necessary to assure that all of the corrosive constituents are removed.

In gas plants handling product streams from the fluid catalytic cracking unit raises the potential for corrosion from moist hydrogen sulfide and cyanide derivatives. When feedstocks are from the visbreaker, the delayed coker, the fluid coker, or any other thermal cracking unit, corrosion from hydrogen sulfide and deposits of iron sulfide in the high-pressure sections of gas compressors from ammonium compounds is possible. Furthermore, processing opportunity crudes requires refiners to manage greater volumes of corrosive and toxic hydrogen sulfide.

Careful considerations in the design and operations of processing options must address the corrosive nature of hydrogen sulfide. In addition, hydrogen sulfide is evolved from all stages of petroleum refining. The major objectives in petroleum refining result in concentrating hydrogen sulfide in different streams, including: (1) processing opportunity high sulfur content feedstocks to improve refining margins, (2) meeting quality specifications for liquefied petroleum gas (LPG), gasoline, low sulfur diesel, and ultralow sulfur diesel, (3) treating feed to limit sulfur content for subsequent processing, (4) hydrocracking product stream to increase distillate yield, and (5) complying with environmental regulations on emissions.

4.2 GAS STREAMS

The actual practice of processing gas streams to pipeline dry gas quality levels can be quite complex, but it usually involves several processes to remove the various impurities. Gas streams produced during petroleum and natural gas refining, while ostensibly being hydrocarbon in nature, may contain large amounts of acid gases, such as hydrogen sulfide and carbon dioxide (Speight and Ozum, 2002; Hsu and Robinson, 2006; Gary et al., 2007; Speight 2007, 2014). First it is necessary to understand the origin of the gas streams and the means by which they may be processed. Gas streams can be conveniently subcategorized into two types of streams: (1) natural gas and (2) refinery gas, also called process gas.

4.2.1 Natural Gas

Natural gas comes from three types of wells: (1) oil wells, (2) gas wells, and (3) condensate wells. Natural gas that comes from oil wells is typically termed *associated gas*. This gas can exist separately from oil in the formation (free gas) or can be dissolved in the crude oil (dissolved gas). Natural gas from a petroleum reservoir will also contain high boiling hydrocarbons as high as C_8. On the other hand, natural gas from gas reservoirs and condensate reservoirs, in which there is little or no crude oil, is termed *nonassociated gas* and generally contains less of the higher boiling hydrocarbons that are part of natural gas from petroleum reservoirs. Gas wells typically produce natural gas only, while condensate wells produce free natural gas along with a semiliquid hydrocarbon condensate.

Whatever be the source of the natural gas, once separated from crude oil (if present), it commonly exists in mixtures with other hydrocarbons: principally ethane, propane, butane, and pentanes. In addition, raw natural gas contains water vapor, hydrogen sulfide (H_2S), carbon dioxide, helium, nitrogen, and other compounds. In fact, associated hydrocarbons, known as *natural gas liquids* (NGLs), can be very valuable by-products of natural gas processing. NGLs include ethane, propane, butane, isobutane, and natural gasoline. They are sold separately and have a variety of different uses, including enhancing oil recovery in oil wells, providing raw materials for oil refineries or petrochemical plants, and as sources of energy.

However, any acidic gases (such as hydrogen sulfide and carbon dioxide) in the gas mixture will corrode refining equipment, pipelines, and storage vessels; harm catalysts; pollute the atmosphere; and prevent the use of hydrocarbon components in petrochemical manufacture. When the amount of hydrogen sulfide is high, it may be removed from a gas stream and converted into sulfur or sulfuric acid. Some natural gases contain sufficient carbon dioxide to warrant recovery as dry ice (Bartoo, 1985; Speight, 2007).

4.2.2 Refinery Gas

Petroleum refining produces gas streams that contain substantial amounts of acid gases, such as hydrogen sulfide and carbon dioxide (Speight and Ozum, 2002; Hsu and Robinson, 2006; Gary et al., 2007; Speight, 2014). The terms *refinery gas* and *process gas* are also often used to include all of the gaseous products and gaseous by-products that emanate from a variety of refinery processes. There are also components of the gaseous products that must be removed prior to release of the gases to the atmosphere or prior to use of the gas in another part of the refinery, for example, as a fuel gas or as a process feedstock.

The processes that evolve hydrogen sulfide as a by-product include crude oil distillation, treating processes to remove sulfides, thermal and catalytic cracking, delayed coking and hydrocracking, and secondary processing, such as hydrotreatment and processes generally known as product improvement processes (Speight and Ozum, 2002; Hsu and Robinson, 2006; Gary et al., 2007; Speight, 2007, 2014). Hydrogen sulfide from sour gas, fuel gas, and LPG streams are removed by using amines (olamines) as a solvent; the hydrogen sulfide from sour water is removed in a sour water stripper. Hydrogen sulfide is then removed in the sulfur recovery unit, which converts hydrogen sulfide into elemental sulfur. All of these processing operations are essential to mitigate the potential for corrosion but also to meet the environmental regulations on air emissions.

Thus, the gas streams are produced during initial distillation of the crude oil and the various conversion processes (Speight, 2007; Sloley, 2013a, b; Speight, 2014). Of particular interest is the hydrogen sulfide (H_2S) that arises from the hydrodesulfurization of feedstocks that contain organic sulfur:

$$[S]_{feedstock} + H_2 \rightarrow H_2S + hydrocarbons$$

Petroleum refining involves, with the exception of heavy crude oil, *primary distillation* (Chapter 3) and results in separation into fractions differing in carbon number, volatility, specific gravity, and other characteristics. The most volatile fraction, which contains most of the gases that are generally dissolved in the crude, is referred to as *pipe still gas* or *pipe still light ends* and consists essentially of hydrocarbon gases ranging from methane(s) to butane(s), or sometimes pentane(s).

The gas varies in composition and volume, depending on crude origin and on any additions to the crude made at the loading point. It is not uncommon to reinject light hydrocarbons, such as propane and butane, into the crude oil before dispatch by tanker or pipeline. This results in a higher vapor pressure of the crude, but it allows one to increase the quantity of low boiling products obtained at the refinery. Since light ends in most petroleum markets command a premium, while in the oil field itself propane and butane may have to be reinjected or flared, the practice of *spiking* crude oil with LPG is becoming fairly common.

In addition to the gases obtained by distillation of petroleum, more highly volatile products result from the subsequent processing of naphtha and middle distillate to produce gasoline (Speight and Ozum, 2002; Hsu and Robinson, 2006; Gary et al., 2007; Speight, 2007, 2014). Hydrogen sulfide is produced in the desulfurization processes, involving hydrogen treatment of naphtha, distillate, and residual fuel; and from the coking or similar thermal treatments of vacuum gas oils and residua. The most common processing step in the production of gasoline is the catalytic reforming of hydrocarbon fractions in the heptane (C_7) to decane (C_{10}) range.

Additional gases are produced in *thermal cracking processes*, such as the coking or visbreaking processes for processing heavy feedstocks. In the visbreaking process, fuel oil is passed through externally fired tubes and undergoes liquid phase cracking reactions, which result in the formation of lighter fuel oil components. Oil viscosity is thereby reduced, and some gases—mainly hydrogen, methane, and ethane—are formed. Substantial quantities of both gas and carbon are also formed in coking (both fluid and delayed coking) in addition to the middle distillate and naphtha. When coking a residual fuel oil or heavy gas oil, the feedstock is preheated and contacted with hot carbon (coke), which causes extensive cracking of the feedstock constituents of higher molecular weight to produce lower molecular weight products ranging from methane, via LPG(s) and naphtha, to gas oil and heating oil. Products from coking processes tend to be unsaturated, and olefin components predominate in the tail gases from coking processes.

Another group of refining operations that contribute to gas production is that of the *catalytic cracking processes*. These consist of fluid-bed catalytic cracking and there are many process variants in which heavy feedstocks are converted into cracked gas, LPG, catalytic naphtha, fuel oil, and coke by contacting the heavy hydrocarbon with the hot catalyst. Both catalytic and thermal cracking processes, the latter being now largely used for the production of chemical raw materials, result in the formation of unsaturated hydrocarbons, particularly ethylene (CH_2═CH_2), but also propylene (propene, CH_3CH═CH_2), *iso*-butylene [*iso*-butene, $(CH_3)_2C$═CH_2], and the *n*-butenes (CH_3CH_2CH═CH_2 and CH_3CH═$CHCH_3$) in addition to hydrogen (H_2), methane (CH_4), and smaller quantities of ethane (CH_3CH_3), propane ($CH_3CH_2CH_3$), and butanes [$CH_3CH_2CH_2CH_3$, $(CH_3)_3CH$]. Diolefins, such as butadiene (CH_2═$CHCH$═CH_2), are also present.

A further source of refinery gas is *hydrocracking*, a catalytic high-pressure pyrolysis process in the presence of fresh and recycled hydrogen. The feedstock is again heavy gas oil or residual fuel oil, and the process is mainly directed at the production of additional middle distillates and gasoline. Since hydrogen is to be recycled, the gases produced in this process again have to be separated into lighter and heavier streams; any surplus recycled gas and the LPG from the hydrocracking process are saturated.

In a series of *reforming processes*, commercialized under names such as *Platforming*, paraffin and naphthene (cyclic nonaromatic) hydrocarbons are converted in the presence of hydrogen and a catalyst into aromatics, or isomerized to more highly branched hydrocarbons. Catalytic reforming processes thus not only result in the formation of a liquid product of higher octane number but also produce substantial quantities of gases. The latter are rich in hydrogen but also contain hydrocarbons from methanes to butanes, with a preponderance of propane ($CH_3CH_2CH_3$), *n*-butane ($CH_3CH_2CH_2CH_3$), and *iso*-butane [$(CH_3)_3CH$].

The composition of the process gas varies in accordance with reforming severity and reformer feedstock. All catalytic reforming processes require substantial recycling of a hydrogen stream. Therefore, it is normal to separate reformer gas into a propane ($CH_3CH_2CH_3$) and/or a butane stream [$CH_3CH_2CH_2CH_3$ plus ($CH_3)_3CH$], which becomes part of the refinery LPG production and a lighter gas fraction, part of which is recycled. In view of the excess of hydrogen in the gas, all products of catalytic reforming are saturated, and there are usually no olefin gases present in either gas stream.

Both hydrocracker and catalytic reformer gases are commonly used in catalytic desulfurization processes. In the latter, feedstocks ranging from light to vacuum gas oils are passed at pressures of $500-1000$ psi with hydrogen over a hydrofining catalyst. This results mainly in the conversion of organic sulfur compounds to hydrogen sulfide:

$$[S]_{feedstock} + H_2 \rightarrow H_2S + \text{hydrocarbons}$$

This process also produces some light hydrocarbons by hydrocracking.

Thus, refinery gas streams, while ostensibly being hydrocarbon in nature, may contain large amounts of acid gases, such as hydrogen sulfide and carbon dioxide. Most commercial plants employ hydrogenation to convert organic sulfur compounds into hydrogen sulfide. Hydrogenation is effected by means of recycled hydrogen-containing gases or external hydrogen over a nickel molybdate or cobalt molybdate catalyst.

The presence of impurities in gas streams may eliminate some of the sweetening processes, since some processes remove large amounts of acid gas, but not to a sufficiently low concentration. On the other hand, there are those processes not designed to remove (or incapable of removing) large amounts of acid gases, whereas they are capable of removing the acid-gas impurities to very low levels when the acid gases are present only in low to medium concentrations in the gas.

The processes that have been developed to accomplish gas purification vary from a simple once-through wash operation to complex multistep recycling systems. In many cases, the process complexities arise because of the need for recovery of the materials used to remove the contaminants or even recovery of the contaminants in the original, or altered, form (Katz, 1959; Newman, 1985; Kohl and Nielsen, 1997). From a corrosion viewpoint, it is not means by which these gases can be utilized, but the effects of these gases on the equipment when they are introduced into reactors for use or into pipelines for transportation.

In addition to the corrosion of equipment of acid gases, the escape into the atmosphere of sulfur-containing gases can eventually lead to the formation of the constituents of acid rain: the oxides of sulfur (SO_2 and SO_3). Similarly, the nitrogen-containing gases can also lead to nitrous and nitric acids (through the formation of the oxides NO_x, where $x = 1$ or 2), which are the other major contributors to acid rain. The release of carbon dioxide and hydrocarbons as constituents of refinery effluents can also influence the behavior and integrity of the ozone layer.

Finally, another acid gas, hydrogen chloride (HCl), although not usually considered to be a major emission, is produced from mineral matter and the brine that often accompanies petroleum during production and is gaining increasing recognition as a contributor to acid rain. However, hydrogen chloride may exert severe local effects because it does not need to participate in any further chemical reaction to become an acid. Under atmospheric conditions that favor a buildup of stack emissions in the areas where hydrogen chloride is produced, the amount of hydrochloric acid in rain water could be quite high.

In summary, refinery process gas, in addition to hydrocarbons, may contain other contaminants, such as carbon oxides (CO_x, where $x = 1$ and/or 2), sulfur oxides (SO_x, where $x = 2$ and/or 3), as well as ammonia (NH_3), mercaptans (R-SH), and carbonyl sulfide (COS). From an environmental viewpoint, petroleum processing can result in a variety of gaseous emissions. It is a question of degree insofar as the composition of the gaseous emissions may vary from process to process, but the constituents are, in the majority of cases, the same.

4.3 CORROSION CHEMISTRY

In sour-gas streams, the two primary corrosion-causing species are hydrogen sulfide (H_2S) and carbon dioxide (CO_2), with contributions from other corrosive constituents. Streams containing ammonia should be dried before processing. Antifouling additives may be used in absorption oil to protect heat exchangers. Corrosion inhibitors may be used to control corrosion in overhead systems.

4.3.1 Hydrogen Sulfide

Iron sulfide (FeS) is the reaction product of iron (Fe) and sulfur (S) in the absence of oxygen. More specifically to the olamine systems of gas processing, the reaction is also the result of the reaction of iron with hydrogen sulfide. This initial reaction is a form of metal corrosion. However, under ideal conditions, the iron sulfide is formed. It then adheres to the walls of the piping and vessel internals and acts as a protective film, thus retarding further metal corrosion. This mechanism is actually one of the main reasons why carbon steel is used in olamine plant construction. Generally, the chemistry of the corrosion caused by hydrogen sulfide in gas streams is represented chemically by the initial formation of iron sulfide.

Corrosion by wet hydrogen sulfide

$$Fe + H_2S \rightarrow FeS + H_2$$

Water is an agent in the onset of corrosion by hydrogen sulfide, which dissolves in water but the solution is very weak; the gas is liberated from the solution with (1) the slightest agitation, (2) reduction in pH, or (3) contact with reactive material.

The reaction mechanism of hydrogen sulfide with steel that results in forming iron sulfide is complex and occurs by several intermediate reactions, but the reaction only takes place in the presence of water. Rust left in the system or pipeline can also lead to FeS formation. Several species or types of FeS—the most commonly found in olamine systems—in order of increasing sulfidation are: (1) mackinawite: FeS_{1-x} or $Fe_{1+x}S$, which is the most soluble type of FeS, as well as most reactive with oxygen; (2) pyrrhotite: $Fe_{1-x}S$, which is iron-deficient sulfide and more stable than mackinawite; and (3) pyrite: FeS_2, which is the most stable form of iron sulfide. Other types of iron sulfide that may be found include greigite (Fe_3S_4), the product of sulfidization of mackinawite, and troilite (FeS), which is the stoichiometric iron sulfide, but which is only rarely found in olamine systems. There are no known field methods to differentiate one FeS species from another; the most common laboratory method is by X-ray diffraction.

Mackinawite is the initial form of iron sulfide that develops in olamine systems and it is fairly soluble, but—as a result—it does not form as strong of a protective layer on the piping wall compared to pyrrhotite or pyrite. The other products are hydrocarbon and polymerized olamine. There can still be some iron sulfide scale on piping and vessel walls because the rate of formation of the iron sulfide is much faster than the dissolution rate back into the olamine. As more and more hydrogen sulfide reacts with the mackinawite, the ratio of sulfur to iron grows and will eventually change the molecular structure of the FeS molecule. With an adequate partial pressure of hydrogen sulfide, mackinawite will convert to pyrrhotite quickly at temperatures above 43°C (109°F). Pyrrhotite is less soluble than mackinawite and deposits on piping walls between 43°C and 150°C (109−302°F), thereby forming a protective (passivating) film. Unless removed, this film prevents further pipe and vessel corrosion and can result in extremely long life spans for that part of the olamine plant. However, that iron sulfide may have formed a strong protective layer in one part of a plant, but not in others.

Pyrite is formed when the ratio of sulfur to iron reaches 2:1. Elemental sulfur can also react with iron to make pyrite, making it common in regenerator bottoms and reboilers. It has extremely low solubility levels and is the hardest type of iron sulfide; pyrite, though very durable, is not a preferred protective film. If even the smallest space exists between it and carbon steel, a galvanic cell (Chapter 1) can be formed between the pipe wall and pyrite that will result in very high corrosion rates.

If hydrogen sulfide is present in the inlet gas, this does not necessarily mean the resulting iron sulfide formed in the system will form a protective layer on the piping walls. The partial pressure of the hydrogen sulfide appears to make a large contribution to the scale depth and quality. However, once the scale is compromised, the chance of developing a galvanic-type corrosion cell is increased. Weaker scales are also easily removed or subject to delamination, allowing for the potential ingress of carbon dioxide under the deposit and the subsequent aggressive under-deposit corrosion that can lead to significant metal failures.

Iron sulfide particles can also enter a facility via the feed gas stream. Ideally, they should be removed by the separation devices at the inlet, but this is not always the case. If iron sulfide is present, it will most likely be removed by the olamine solution (which acts like a water-wash column, which is another way to remove iron sulfide upstream of a process). In these cases, the iron sulfide will simply add to the system suspended solids content and can cause a number of problems, such as plugging and flow disturbance.

In many instances, olamine plants suffer from iron sulfide intrusion through being carried by the inlet gas. In such cases, the iron sulfide particles will not react with the piping walls and add to the protective film, but they will remain as suspended solids. As a result, the suspended solids will scour off the previously formed protective films that contribute to a greater quantity of suspended solids in solution, as well as to an erratic corrosion protection film within the system, leaving freshly exposed metal surfaces as active sites for the corrosion mechanism to continue.

Hydrogen sulfide corrosion results in the formation of black iron sulfide scales and is typified by *black water* in the separation units. Under-deposit corrosion frequently occurs beneath the scale layer and can result in forming deep, isolated, or randomly scattered pits. The three prime means of removing or reducing the impact of iron sulfide entering an olamine system are to: (1) prevent the corrosion from occurring initially in the piping by using corrosion inhibitors, (2) disperse the iron sulfide particles into the water phase so they can be removed by inlet separation equipment, and (3) remove the iron sulfide from the gas phase upstream of the olamine absorber by use of a suitable filter or by a water wash.

Iron sulfide can be formed in the absorber, piping system, or reboiler/regenerator. In the absorber, soluble iron is present in lean-olamine streams. The iron may be in the form of iron carbonate ($FeCO_3$) in instances where carbon dioxide is being treated as well as hydrogen sulfide. In the absorber, some of the hydrogen sulfide immediately reacts with iron in the olamine, and small iron sulfide particles are formed. These particles are generally insoluble in the olamine and, provided they are large enough, can be removed by a filter.

Both the rich and lean olamine will have hydrogen sulfide in solution and the rich olamine obviously will have a much higher amount. Higher partial pressures of hydrogen sulfide result in higher tendencies for strong iron sulfide films to form. However, when the partial pressure of hydrogen sulfide is low, the resulting iron sulfide is typically mackinawite. Mackinawite does not form a strong adhesive protective layer on the piping; instead, it is preferentially carried by the solution and moves along with the olamine, resulting in lean/rich exchanger plugging as well as other associated problems.

Iron sulfide films are stronger and thicker in plant areas where the partial pressure of hydrogen sulfide is the highest—typically in the pipe carrying rich olamine from the absorber to the flash tank. As system pressure or hydrogen sulfide content in the olamine decreases, the iron sulfide film decreases in thickness and quality. At the same time, because there is low partial pressure of hydrogen sulfide, there is generally less need for protection, provided there is no significant carbon dioxide content or aggressive organic acid level in either the solution or the vapor phase.

In the regenerator tower lower section and in the reboiler, the partial pressure of the hydrogen sulfide is extremely low and iron sulfide formation is minimal. Elemental sulfur, however, which enters a plant bonded with hydrogen sulfide as hydrogen polysulfide (H_2S_x), is liberated when the hydrogen sulfide is driven off and is no longer soluble in the olamine solution. Elemental sulfur reacts quickly with iron to form pyrite (FeS_2), which is the predominant scale found in this area. However, if the hydrogen sulfide remains in solution, the usual iron sulfide reaction will still occur but—due to the high temperatures driving the reaction—the iron sulfide formed will be pyrrhotite or pyrite.

As indicated above, it is generally desirable to leave the iron sulfide film on the olamine plant internals. This film can be removed accidentally such as by: (1) high fluid velocity, (2) excessive vibration, (3) mechanical/ thermal shocks during start-up/shut-down, (4) heat-stable salt degradation products (increased suspended solids erode the FeS layer), (5) chelating agents present in the liquid phase, and (6) adding a corrosion inhibitor to the system without understanding the protection mechanism.

Once liberated, the suspended iron sulfide particles can result in several problems: (1) olamine foaming, which results in off-specification gas and the tendency to carry over rather than *cause* a solution to foam, solids tend to *stabilize* an already foaming condition, (2) excessive mechanical wear on pumps and seals; lost efficiency and higher maintenance frequency, (3) lost olamine efficiency, which curtails throughput, (4) higher chemical use/costs, (5) abrasion, in which the suspended iron sulfide erodes the existing iron sulfide film in other areas, (6) excessive particle filter plugging and usage, and (7) packing, tray valve, or sieve hole plugging.

High fluid velocity—the iron sulfide particles that have formed between hydrogen sulfide in the liquid phase and iron in the piping walls may be under such high drag forces that they cannot adhere to the piping walls. With no protective film, fresh iron is exposed that will also react with hydrogen sulfide. In cases where olamine circulation rate is to be significantly increased, it is recommended that pipe internal diameters are double-checked to ensure the fluid velocities do not get too high.

Excessive vibration—strong iron sulfide films can crack and break loose if the pipe begins to vibrate excessively, which is often associated with high heat flux in the reboiler tubes.

Start-up/shut-down shocks—it is common for filter plugging to occur immediately after starting an olamine system (commonly referred to as an *upset*) and, much like excessive vibration, sudden surges and thermal shocks displace the iron sulfide layer from the pipe walls. The particles knocked loose are large and easily picked up by filters but can plug lines to other pieces of equipment.

Increase in heat-stable olamine salt levels—the solubility of iron sulfide (pyrite being the exception) in the olamine solution increases as the solution becomes more acidic that is due, for example, to: (1) an increase in hydrogen sulfide and carbon dioxide concentration, (2) change in olamine type or strength, or (3) heat-stable salt buildup. Iron sulfide formed at pH levels below 8.5 are much less effective at adhering strongly to the pipe walls. The negative effects of high acidity (low pH) are found most predominantly in high-temperature areas, as the pH drops, the existing iron sulfide film softens as anions from the acid react with the iron and, if allowed to continue, the iron sulfide film is eventually removed.

Chelating agents present—iron complexing agents (chelating agents), such as cyanide, thiocyanate, and ethylenediaminetetraacetic acid (EDTA), will act to solubilize the otherwise insoluble iron sulfide into solution. Olamines are capable of holding much more iron if chelating agents are in the solution, but without chelating agents, the olamine can typically hold less than 5 ppm iron.

In addition, the effluent streams from any of the refining processes can contain ammonia (NH_3) and hydrogen sulfide (H_2S), which react to form ammonium bisulfide (NH_4HS), which is highly corrosive to carbon steel and may lead to a catastrophic failure. The severity of ammonium bisulfide-induced corrosion depends upon: (1) the concentration of ammonium bisulfide, (2) the fluid velocity and turbulence, (3) wash-water management, as well as (4) piping configuration and temperature of the system. The areas that are most vulnerable to ammonium bisulfide corrosion are: (1) the reactor-effluent air coolers (REACs) and (2) the upstream and downstream piping.

The precipitation of the ammonium bisulfide on effluent condenser tubes leads to cause fouling, under-deposit corrosion, and tube pitting. The fouling and deposits increase localized fluid velocity, thus supporting a higher localized corrosion rate. Corrosion is characterized by localized spiral gouging in a straight portion of piping and severe metal loss at the elbows and inlet ends of coolers. The protective surface film is washed off by a high-velocity process stream, thus exposing a fresh metal surface for attack. Contaminants such as chlorides, oxygen, and cyanides also aggravate the corrosion rate. The system temperature plays an important role in the precipitation of salt and impacts where wash-water injection point locates.

The corrosion mitigation procedure involves monitoring wash-water quality and the quantity needed to dissolve the salts. Approximately 25% of wash water should remain in the aqueous phase at the point of injection. Good distribution of wash water in the system and low ammonium bisulfide concentration in the separator water should be targeted. The system should operate in the fluid-velocity range to avoid phase separation at low velocity and corrosion at high velocity.

Carbon steel metallurgy is suitable under listed processing conditions. However, deviations due to reactor-feed quality or unit augmentation may require using inhibitors or a metallurgy upgrade. Filming-amine inhibitors have been effective in controlling moderate corrosion conditions. However, a metallurgy upgrade may be necessary under severe corrosion conditions.

4.3.2 Carbon Dioxide

Corrosion by wet carbon dioxide corrosion can result in high corrosion rates, but a carbonate film gives some protection and is more protective at higher temperatures. The carbon dioxide content is often not very high in refinery streams, except in hydrogen reformer plant systems. In addition, one of the major sources of corrosion on carbon steel vessels in sweetening units is heat-stable material, which is a product of amine degradation (Rooney and Dupart, 2000). Oxygen plays a major role in amine degradation—the reaction of oxygen and the amine produces organic acids, such as acetic acid and formic acid.

Corrosion in the amine unit (especially in the presence of rich amine solutions) is increased by high acid-gas loading—thus, the loading often has to be limited to minimize corrosion. Acid-gas flashing disturbs the protective film of iron sulfide (FeS) protective films. Acid gases break out of solution to give acid attack when there is a high velocity and high temperature and when the pressure is too low to suppress vaporization. It is also important to avoid too low a level of hydrogen sulfide in the lean amine—a small amount of hydrogen sulfide is helpful in producing a protective sulfide film. Primary amines (RNH_2) are more corrosive than secondary amines (R^1R^2NH) and tertiary amines ($R^1R^2R^3N$). Hydrogen sulfide forms protective sulfide films on carbon steel in many areas, but there are problems in areas where films form and then can be removed. In such locations, upgrading of materials is required, often to an austenitic stainless steel belonging to the 300 series.

Heat-stable olamine salts form from stronger acids than hydrogen sulfide and carbon dioxide and they do not thermally break down at regeneration temperatures. Problems arise from formic acid, oxalic acid, and acetic acid, as well as from thiosulfurous acids and from chlorides, sulfates, and thiosulfates. Oxygen is also a source of problems and this can come in from the feedstock, olamine storage, and makeup water. Blanketing tanks with nitrogen and maintaining a closed system are helpful in order to exclude oxygen. High temperatures are also a problem and temperatures should be minimized through control of the reboiler temperature.

Heat-stable olamine salts can also be produced from carbon monoxide and hydrogen cyanide. It may be opportune to treat such gas streams, especially streams from fluid catalytic cracking units, with polysulfide inhibitors to remove hydrogen cyanide. The presence of heat-stable olamine salts reduces acid-gas removal capacity, lowers pH, increases conductivity, and dissolves protective films so heat-stable olamine salts should be minimized as much as possible. Makeup water should ideally have low total dissolved solids and low total hardness owing to calcium, low chlorides, sodium, potassium, and dissolved iron, and should exclude oxygen.

Erosion corrosion (Chapter 1) is caused by rich olamine solutions that contain particulate matter—therefore lean olamine is filtered to minimize solids. Protective films of iron sulfide can be damaged and removed under conditions of high velocity, turbulence, or impingement. Benefit can, therefore, be obtained by designing to minimize impingement and turbulence, for example, by using large radius bends. The velocity in piping is usually kept below 1 m/s and 300 series stainless steel is required at pressure letdown valves.

4.3.3 Other Corrosive Agents

Corrosion in the overheads of the regenerator unit takes a different form from that occurring elsewhere in the olamine unit. Hydrogen sulfide, ammonia, and hydrogen cyanide are important species that are involved, which can give corrosion. Conditions are more aggressive when treating streams from cokers, visbreakers, fluid catalytic cracking units, and hydroprocessing units. Ammonium hydrosulfide (NH_4HS) can be particularly aggressive, and close attention needs to be paid to concentration and velocity with this species.

Hydrogen cyanide is detrimental as it removes sulfide scales, which increases corrosion and promotes hydrogen pickup and damage:

$$FeS + 6CN^- \rightarrow Fe(CN)^{6-} + S^{2-}$$

Special attention is needed in order to avoid excessive accumulation of ammonium hydrosulfide and hydrogen cyanide in the regenerator overhead reflux system.

Sulfide stress cracking is prevented by minimizing the hardness and strength of the alloys used for wet hydrogen sulfide systems. This is accomplished through material selection, and the control of weld procedures and postweld heat treatment. Hydrogen pressure–induced cracking, including stress-oriented, hydrogen-induced cracking, is mitigated by the use of improved-quality steel plate and postweld heat treatment or the use of corrosion-resistant alloy cladding (the application of one material over another to provide a skin or layer intended to control corrosion). Carbon steel can be used with success for many areas, but material upgrading is necessary in highly corrosive areas. Use has been made of materials such as the austenitic stainless steels 304 and 3161, 2205 duplex stainless steel, and other high-alloy materials, such as Alloy C or Stellite alloy (cobalt–chromium alloy) for valve trim.

4.4 GAS CLEANING

The actual practice of processing gas streams to pipeline gas quality levels can be quite complex but usually involves four main processes to remove the various impurities. Gas streams produced during petroleum and natural gas refining, while ostensibly being hydrocarbon in nature, may contain large amounts of acid gases such as hydrogen sulfide and carbon dioxide. Most commercial plants employ hydrogenation to convert organic sulfur compounds into hydrogen sulfide. Hydrogenation is effected by means of recycled hydrogen-containing gases or external hydrogen over a nickel molybdate or cobalt molybdate catalyst. Also, process gas, in addition to hydrocarbons, may contain other contaminants, such as carbon oxides (CO_x, where $x = 1$ and/or 2), sulfur oxides (SO_x, where $x = 2$ and/or 3), as well as ammonia (NH_3), mercaptans (R-SH), and carbonyl sulfide (COS).

The presence of these impurities may eliminate some of the sweetening processes, since some processes remove large amounts of acid gas, but not to a sufficiently low concentration. On the other hand, there are those processes not designed to remove (or incapable of removing) large amounts of acid gases, whereas they are capable of removing the acid-gas impurities to very low levels when the acid gases are present only in low-to-medium concentrations in the gas (Katz, 1959).

4.4.1 Acid-Gas Removal
In sour-gas production, the two primary corrosion-causing species are hydrogen sulfide (H_2S) and carbon dioxide (CO_2). The corrosion products that form from the reaction of these gases with steel in the presence of water can provide an indication of: (1) the formation mechanism, (2) the severity of the potential corrosive environment, and (3) the degree to which the corrosion will affect operation of the amine unit. In any amine system, piping and equipment corrosion is one of the worst of the potential problems an operator or engineer can encounter. Furthermore, amine plants treating gas containing hydrogen sulfide will have iron sulfides in the system.

Briefly, the simple name *amine* is used to convey, in this case, a more complex molecule. The *amine* is in reality an *olamine*, a bifunctional molecule containing the olamine function ($-NH_2$) and an alcohol function ($-OH$). The simplest member of this family is ethanolamine ($HOCH_2CH_2NH_2$), also known as monoethanolamine (MEA). Other more complex olamine systems are also known for their use in gas cleaning (Table 4.1) and are selected depending upon the properties of the gas stream and the extent of the cleaning required.

The olamine unit (Figure 4.1) is a key component of the gas processing plant, and the olamine regeneration tower is one of the major parts of this unit. Olamine towers in gas plants worldwide have experienced severe corrosion, and the corrosion area is usually (but not always) limited to the vapor and vapor/liquid interface spaces. The olamine itself is not corrosive, but corrosion is promoted by the following: (1) entrained acid gases, (2) higher concentration of corrosive species, (3) higher temperatures, and (4) corrosion on heat transfer surfaces. In addition, olamine units often use a knockout pot before the absorber where liquid hydrocarbon and water are removed, thereby removing some of the potential for inefficient unit operation and corrosion.

The olamine system removes the hydrogen sulfide from these streams: fuel gas from crude distillation and delayed coking units, sponge absorber gas of the fluid catalytic cracking unit, and off-gas from the hydrotreater and hydrocracker units. This is typically accomplished by contacting the sour gas with a 25% w/w solution of diethanolamine (DEA) in an absorption tower (Mokhatab et al., 2006; Speight, 2007, 2014).

The rich DEA (rich in hydrogen sulfide) from the olamine absorption column is routed to the olamine regenerator to strip out the hydrogen sulfide and carbon dioxide. The stripped gas is sent to the sulfur recovery unit. The sweet gas leaving the top of the olamine absorption column is routed to the refinery's internal fuel gas network through the sweet-gas knockout drum. Condensate from the sweet-gas knockout drum is recycled to the olamine regenerator along with its feed. The stripped gas from the olamine stripper and rich olamine are potential risk sources.

Oil from tight shale formations typically has high hydrogen sulfide loading. To ensure worker safety, scavengers are often used to reduce

Table 4.1 Olamines Used for Gas Processing

Olamine	Formula	Derived Name	Molecular Weight	Specific Gravity	Melting Point, °C	Boiling Point, °C	Flash Point, °C	Relative Capacity, %
Ethanolamine (monoethanolamine)	$HOC_2H_4NH_2$	MEA	61.08	1.01	10	170	85	100
Diethanolamine	$(HOC_2H_4)_2NH$	DEA	105.14	1.097	27	217	169	58
Triethanolamine	$(HOC_2H_4)_3NH$	TEA	148.19	1.124	18	335, d	185	41
Diglycolamine (hydroxyethanolamine)	$H(OC_2H_4)_2NH_2$	DGA	105.14	1.057	−11	223	127	58
Diisopropanolamine	$(HOC_3H_6)_2NH$	DIPA	133.19	0.99	42	248	127	46
Methyldiethanolamine	$(HOC_2H_4)_2NCH_3$	MDEA	119.17	1.03	−21	247	127	51

d, with decomposition.

hydrogen sulfide concentrations. The scavengers are often olamine-based products—methyl triazine, for example—that are converted into MEA in the crude distillation unit. Unfortunately, these olamines contribute to corrosion problems in the crude distillation unit. Once MEA forms, there is a rapid reaction with chlorine or hydrogen chloride to form chloride salts. These salts lose solubility in the hydrocarbon phase and become solids at the processing temperatures of the atmospheric distillation tower and form deposits on the trays or overhead system. The deposits are hygroscopic and, once water is absorbed, the deposits become very corrosive.

4.4.2 Sour Water Stripper

The sour water stripper removes hydrogen sulfide and ammonia from sour-water streams from the crude and vacuum units, delayed coker, resid fluid catalytic cracking unit, diesel hydrodesulfurization/hydrotreating unit, and hydrocracker. Sour waters from all of these units are collected in a degassing drum where some light hydrocarbons and hydrogen sulfide are vented to the acid-gas flare. After degassing, the sour water is routed to the sour water stripper. Ammonia and hydrogen sulfide are driven from the water by heat and stripped off by the rising steam/water vapor. Off-gas is sent to the sulfur recovery unit. The stripped water is removed from the bottom of the stripper. Part of

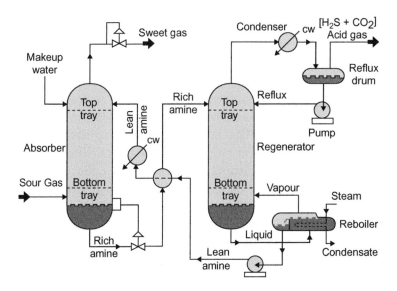

Figure 4.1 Schematic of an olamine unit (amine unit) for gas cleaning.

the water is recoiled in a thermo-siphon steam reboiler. All of the off-gas lines that carry hydrogen sulfide are steam jacked to prevent hydrogen sulfide—rich gas streams from leaking into the atmosphere. The steam jacket also serves to prevent excessive condensation of water, which could dissolve the acid gases and lead to rapid corrosion problems. Sour water to the stripper and stripped gas has a high potential for causing corrosion.

For the sour water stripper, the feed is mainly from units where hydrogen sulfide and ammonia are present. The column-top temperature of the sour water stripper is kept at around 85°C (185°F) to prevent deposition of ammonium sulfide and polysulfide, which otherwise results in plugging and corrosion problems. The temperature at the top of the column should ensure effective rejection of the hydrogen sulfide overhead to the off-gas and minimize acid-gas concentration at the top section of the column. The off-gas line from the degassing drum and stripper are kept hot by steam jacket to prevent condensation and salt formation.

As a rule of thumb, hydrogen sulfide and ammonia are considered tolerable for carbon steel if (%mole hydrogen sulfide) $\times$ (%mole NH_3) is less than 0.5. This is the K_p factor of the sour water stripper to determine its corrosivity. The overhead piping is normally fabricated using carbon steel metallurgy. Monitoring the hydrogen sulfide and ammonia concentrations at the overhead and inspecting the line are necessary to prevent accidental failure.

4.4.3 Sulfur Recovery Unit

The feed gas to sulfur recovery unit is a mixture of acid gas from the olamine recovery unit and sour gas from the sour water stripper. The acid gas from the olamine recovery unit introduced via the olamine-acid-gas knockout drum. The acid gas from sour water stripper is introduced via the sour water stripper gas knockout drum. Sour water separated in the knockout drums is intermittently collected in a sour-water drain vessel and routed back to the olamine recovery unit or sour water stripper unit by nitrogen propulsion. The acid gas is heated to at least 92°C (180°F) in a low-pressure steam-jacketed line to prevent deposition of ammonium salts and/or condensation of water vapor when it is mixed with sour water stripper gas. A mixing temperature of at least 71°C (160°F) is essential. The mixed gas is further heated to 200°C (390°F) in the acid-gas preheater.

The sulfur recovery unit operation consists of the main burner and combustion chamber, Claus reactors, and the sulfur stripping/degassing system. In the main burner and combustion chamber, the air supplied is exactly sufficient to accomplish the complete oxidation of all hydrocarbons and ammonia present in the feed gas and to burn as much hydrogen sulfide as required to obtain a gas stream, which at the outlet of the third reactor, contains 0.5−0.7% v/v hydrogen sulfide. In the first, second, and third reactor stages, the hydrogen sulfide and sulfur dioxide (SO_2) react again to form sulfur, but at a lower temperature and with a catalyst. In the Super-Claus stage, the remaining hydrogen sulfide is selectively oxidized to sulfur. The produced sulfur contains hydrogen sulfide, partly dissolved and partly present in the form of polysulfides (H_2S_x). Without sulfur treatment, the hydrogen sulfide would be slowly released during storage and transport and could create an explosive mixture due to the lower explosive limit of hydrogen sulfide in air, which may vary from 3.7% v/v at 130°C (265°F) to 4.3% v/v at ambient conditions.

The sulfur degassing process degasifies liquid sulfur to 10 ppm w/w hydrogen sulfide, which is the safe level to avoid exceeding the lower explosive limit. The gases to the sulfur recovery unit and sulfur locks are potential risk sources.

4.4.3.1 Claus Process

The disposition of hydrogen sulfide, a toxic gas that originates in crude oils and is also produced in the coking, catalytic cracking, hydrotreating, and hydrocracking processes, is an issue with many refiners. Burning hydrogen sulfide as a fuel gas component or as a flare gas component is precluded by safety and environmental considerations since one of the combustion products is the highly toxic sulfur dioxide (SO_2), which is also toxic. As described above, hydrogen sulfide is typically removed from the refinery light ends gas streams through an olamine process after which application of heat regenerates the olamine and forms an acid-gas stream. Following from this, the acid-gas stream is treated to convert the hydrogen sulfide elemental sulfur and water. The conversion process utilized in most modern refineries is the Claus process, or a variant thereof.

The Claus process involves combustion of approximately one-third of the hydrogen sulfide to sulfur dioxide and then reaction of the sulfur dioxide with the remaining hydrogen sulfide in the presence of a fixed bed of activated alumina, a cobalt molybdenum catalyst, resulting in the formation of elemental sulfur:

$$2H_2S + 3O_2 \rightarrow 2SO_2 + 2H_2O$$
$$2H_2S + SO_2 \rightarrow 3S + 2H_2O$$

Different process flow configurations are in use to achieve the correct hydrogen sulfide/sulfur dioxide ratio in the conversion reactors.

In a split-flow configuration, one-third split of the acid-gas stream is completely combusted and the combustion products are then combined with the noncombusted acid-gas upstream of the conversion reactors. In a once-through configuration, the acid-gas stream is partially combusted by only providing sufficient oxygen in the combustion chamber to combust one-third of the acid gas. Two or three conversion reactors may be required depending on the level of hydrogen sulfide conversion required. Each additional stage provides incrementally less conversion than the previous stage.

Overall, conversion of 96–97% of the hydrogen sulfide to elemental sulfur is achievable in a Claus process. If this is insufficient to meet air quality regulations, a Claus process tail gas treating unit is utilized to remove essentially the entire remaining hydrogen sulfide in the tail gas from the Claus unit. The tail gas treating unit may employ a proprietary solution to absorb the hydrogen sulfide followed by conversion to elemental sulfur.

4.4.3.2 Other Processes
The SCOT (Shell Claus Off-gas Treating) unit is a tail gas unit and uses a hydrotreating reactor followed by olamine scrubbing to recover and recycle sulfur, in the form of hydrogen, to the Claus unit.

In the process, tail gas (containing hydrogen sulfide and sulfur dioxide) is contacted with hydrogen and reduced in a hydrotreating reactor to form hydrogen sulfide and water. The catalyst is typically cobalt/molybdenum on alumina. The gas is then cooled in a water contractor. The hydrogen sulfide–containing gas enters an olamine absorber that is typically in a system segregated from the other refinery olamine systems. The purpose of segregation is twofold: (1) the tail gas treating unit frequently uses a different olamine than the rest of the plant and (2) the tail gas is frequently cleaner than the refinery fuel gas (in regard to contaminants) and segregation of the systems reduces maintenance requirements for the SCOT® unit. Olamines chosen for use in the tail gas system tend to be more selective for hydrogen sulfide and are not affected by the high levels of carbon dioxide in the off-gas.

The hydrotreating reactor converts sulfur dioxide in the off-gas to hydrogen sulfide that is then contacted with a Stretford solution (a mixture of a vanadium salt, anthraquinone disulfonic acid, sodium carbonate, and sodium hydroxide) in a liquid–gas absorber. The hydrogen sulfide reacts stepwise with sodium carbonate and the anthraquinone sulfonic acid to produce elemental sulfur, with vanadium serving as a catalyst. The solution proceeds to a tank where oxygen is added to regenerate the reactants. One or more froth or slurry tanks are used to skim the product sulfur from the solution, which is recirculated to the absorber.

Other tail gas treating processes include: (1) caustic scrubbing, (2) polyethylene glycol treatment, (3) Selectox process, and (4) sulfite/bisulfite tail gas treating (Mokhatab et al., 2006; Speight, 2007, 2014).

Removal of larger amounts of hydrogen sulfide from gas streams requires a continuous process, such as the *Ferrox* process or the Stretford process. The *Ferrox process* is based on the same chemistry as the iron oxide process except that it is fluid and continuous. The *Stretford* process employs a solution containing vanadium salts and anthraquinone disulfonic acid. Most hydrogen sulfide removal processes return the hydrogen sulfide unchanged, but if the quantity involved does not justify installation of a sulfur recovery plant (usually a Claus plant), it is necessary to select a process that directly produces elemental sulfur.

4.4.4 Acid-Flare Header

This header collects all the bleeds and the emergency discharge streams from the pressure safety valves of the sulfur block: the sulfur recovery unit, the sour water stripper, and the olamine recovery unit. The gas enters the sour flare knockout drum where liquid is separated from the gas prior to being sent to the flare stack. The separated liquid (mainly sour water) is sent to the sour water stripper for processing. From the flare KO drum, the gas is sent to the acid-gas stack. The drain from the KO drum is the potential risk area.

REFERENCES

Bartoo, R.K., 1985. In: Newman, S.A. (Ed.), Acid and Sour Gas Treating Processes. Gulf Publishing, Houston, TX.

Gary, J.G., Handwerk, G.E., Kaiser, M.J., 2007. Petroleum Refining: Technology and Economics, fifth ed. CRC Press, Taylor & Francis Group, Boca Raton, FL.

Hsu, C.S., Robinson, P.R. (Eds.), 2006. Practical Advances in Petroleum Processing Volume 1 and Volume 2. Springer Science, New York, NY.

Katz, D.K., 1959. Handbook of Natural Gas Engineering. McGraw-Hill Book Company, New York, NY.

Kohl, A.L., Nielsen, R.B., 1997. Gas Purification. Gulf Publishing Company, Houston, TX.

Mokhatab, S., Poe, W.A., Speight, J.G., 2006. Handbook of Natural Gas Transmission and Processing. Elsevier, Amsterdam, The Netherlands.

Newman, S.A., 1985. Acid and Sour Gas Treating Processes. Gulf Publishing, Houston, TX.

Rooney, C.P., Dupart, M.S., 2000. Corrosion in alkanolamine plants: causes and minimization. In: Proceedings of the Corrosion 2000, NACE International. Orlando, FL.

Sloley, A.W., 2013a. Mitigate fouling in crude unit overhead Part 1. Hydrocarb. Process. 92 (9).

Sloley, A.W., 2013b. Mitigate fouling in crude unit overhead Part 2. Hydrocarb. Process. 92 (11), 73−75.

Speight, J.G., 2007. Natural Gas: A Basic Handbook. GPC Books, Gulf Publishing Company, Houston, TX.

Speight, J.G., 2014. The Chemistry and Technology of Petroleum, fifth ed. CRC Press, Taylor & Francis Group, Boca Raton, FL.

Speight, J.G., Ozum, B., 2002. Petroleum Refining Processes. Marcel Dekker Inc., New York, NY.

FURTHER READING

Ajeel, S.A., Abdullatef Ahmed, M.A., 2008. Study of the synergy effect on erosion−corrosion in oil pipes. Eng. Tech. 26 (9), 1068.

Barbouteau, L., Dalaud, R., 1972. In: Nonhebel, G. (Ed.), Gas Purification Processes for Air Pollution Control. Butterworth and Co., London, UK (Chapter 7).

Barnett, J.W., 1988. Desalters can remove more than salts and sediment. Oil Gas J. April, 43.

Chambers, B., Srinivasan, S., Yap, K.M., Yunovich, M., 2011. Corrosion in crude distillation unit overhead operations: a comprehensive review. In: Proceedings of the Corrosion 2011, NACE International. 13−17 March, Houston, TX.

Choi, D.W., 2005. How to buy and operate desalters. Hydrocarb. Process. March, 75.

Curry, R.N., 1981. Fundamentals of Natural Gas Conditioning. PennWell Publishing Co., Tulsa, OK.

Dos Santos Liporace, F., de Oliveira, S.G., 2005. Real time fouling diagnosis and heat exchanger performance. In: Müller-Steinhagen, H., Malayeri, M.R., Watkinson, A.P. (Eds.), Proceedings of the ECI Symposium Series, Volume RP2. Sixth International Conference on Heat Exchanger Fouling and Cleaning—Challenges and Opportunities. Engineering Conferences International, Kloster Irsee, Germany, June 5−10.

Fulker, R.D., 1972. In: Nonhebel, G. (Ed.), Gas Purification Processes for Air Pollution Control. Butterworth and Co., London, UK (Chapter 9).

Gutzeit, J., 2000. Effect of organic chloride contamination of crude oil on refinery corrosion. In: Proceedings of the Corrosion/2000. NACE International, Houston, TX.

Gutzeit, J., 2008. Controlling crude unit overhead corrosion by improved desalting. Hydrocarb. Process. February, 119.

Jambo, H.C.M., Freitas, D.S., Ponciano, J.A.C., 2002. Ammonium hydroxide injection for overhead corrosion control in a crude distillation unit. In: Proceedings of the International Corrosion Congress, Granada, Spain, September.

Jou, F.Y., Otto, F.D., Mather, A.E., 1985. Solubility of H_2S and CO_2 in triethanolamine solutions (Chapter 10). In: Newman, S.A. (Ed.), Acid and Sour Gas Treating Processes. Gulf Publishing Co., Houston, pp. 278–288.

Kane, R.D., 2006. Corrosion in petroleum refining and petrochemical operations. Corrosion: environments and industries. ASM Handbook Volume 13C. ASM International, Materials Park, OH, pp. 967–1014.

Kronenberger, D.L., 1984. Paper No. 128. Corrosion problems associated with the desalting difficulties of Maya and Other Heavy Crudes. In: Proceedings of the Corrosion/84. NACE International, Houston, TX.

Kunnas, J., Ovaskainen, O., Respini, M., 2010. Mitigate fouling in ebullated-bed hydrocrackers. Hydrocarb. Process. 10 (1).

Lachet, V., Béhar, E., 2000. Industrial perspective on natural gas hydrates. Oil Gas Sci. Technol. 55, 611–616.

Lindemuth, P.M., Lessard, R.B., Lozynski, M., 2001. Improve desalter operations. Hydrocarb. Process. September, 67.

Mandal, K.K., 2005. Improve desalter control. Hydrocarb. Process. April, 77.

Nonhebel, G., 1964. Gas Purification Processes. George Newnes Ltd., London, UK.

Speight, J.G., 2000. The Desulfurization of Heavy Oils and Residua, second ed. Marcel Dekker Inc., New York, NY.

Speight, J.G., 2011. An Introduction to Petroleum Technology, Economics, and Politics. Scrivener Publishing, Salem, MA.

Stark, J.L., Asomaning, S., 2003. Crude oil blending effects on asphaltene stability in refinery fouling. Pet. Sci. Technol. 21 (3 & 4), 569–579.

Ward, E.R., 1972. In: Nonhebel, G. (Ed.), Gas Purification Processes for Air Pollution Control. Butterworth and Co., London, UK (Chapter 8).

Zhang, L.Q., Shi, L.B., Zhou, Y., 2007. Formation prediction and prevention technology of natural gas hydrate. Nat. Gas Technol. 1 (6), 67–69.

CHAPTER 5

Corrosion in Other Systems

5.1 INTRODUCTION

Corrosion is a natural phenomenon (Chapter 1) that occurs not only in refineries and gas processing plants but in many other related areas associated with petroleum refining and gas processing. The equipment and components that comprise the related operations must also be fabricated from durable alloys and be suitable to withstand the harsh conditions of the refining industry—stainless steel is one such material (White and Ehmke, 1991; Garverick, 1994).

It is the purpose of this chapter to present the corrosion that occurs in the related operations as well as the key aspects of corrosion processes and the types of corrosion that can occur.

5.2 HEAT EXCHANGERS

Corrosive components in feed streams, as well as in complete refinery processes, are a major cause of heat-exchanger tube failure. These types of failures are usually related to a complex deposition mechanism formation and growth. Deposit layer thickening hinders heat transfer and obstructs pipe.

The problems associated with heat-exchanger fouling have been known since the invention of the heat exchanger. Despite the best efforts of engineers and technologists to reduce or eliminate heat-exchanger fouling, deposit growth still occurs in some cases. Periodic heat-exchanger cleaning is necessary to restore the heat exchanger to efficient operation. Scheduled and unscheduled shutdowns for cleaning can be very expensive because the unit startups may be very time consuming. Thus, anything that can be done to reduce these shutdowns along with the cleaning procedure is of great benefit. When corrosion combines with fouling, the problem is more serious and complicated.

Material selection for process equipment construction has a significant impact on plant efficiency. Among available metals and alloys, a few can be used for process equipment and piping construction. Carbon steel is used for most components in refineries because it is inexpensive and readily available. It can cause a lot of corrosion-related problems, however, which are dependent upon the process unit, the process parameters, and the mechanism of corrosion (Grabke et al., 1995; Stott and Shih, 2000; Asteman and Spiegel, 2007).

Two hypotheses may be proposed: (1) local corrosion followed by organic deposition on the corrosion particle surface and (2) corrosion under the deposit. It is possible that the deposit accelerates heat-exchanger corrosion. Some corrosion fouling is initiated by surface particles subjected to corrosion. To understand the mechanism exactly, several experiments should be done. Streams, deposits, and also tubes should be analyzed to precisely reveal the failure mechanism.

The most probable mechanism is overhead acid corrosion, and the hypothesis of local corrosion and organic deposition cannot be viable because the tube bundle would be disordered by so many corrosion products. The corrosion products settle due to high residence time in the heat exchanger.

Drum water drainage data showed the amount of chlorides to be in the range of 120–440 ppm, supporting the mechanism discussed previously. The chloride content is composed of sodium, calcium, and magnesium chloride. Magnesium and calcium chloride start to hydrolyze at 120°C and 220°C (250°F and 430°F), respectively. Sodium chloride does not hydrolyze at such normal temperatures in the reboiler:

$$CaCl_2 + 2H_2O \rightarrow Ca(OH)_2 + 2HCl$$
$$MgCl_2 + 2H_2O \rightarrow Mg(OH)_2 + 2HCl$$

The reboiler tower outlet temperature is approximately 230°C (445°F) and, therefore, the evolved hydrochloric acid is distilled up to the overhead system. The initial condensate forms after the vapor—containing a high percentage of hydrogen chloride—leaves the column. Due to the high acid concentration dissolved in the water, the pH of the first condensate is quite low. Therefore, at dew point, water condensate corrodes the low-carbon-steel air cooler. Corrosion by acidic chloride condensates is driven by the hydrogen ion concentration (pH) via the reaction:

$$Fe + 2HCl \rightarrow FeCl2 + H_2$$

Although the corrosive attack source is hydrogen chloride, the corrosion product is iron sulfide, not iron chloride. Iron sulfide is precipitated by the reaction between hydrogen sulfide and soluble iron chlorides from the corrosion reaction between hydrogen chloride and iron:

$$FeCl_2 + H_2S \rightarrow FeS_2 + 2HCl$$

The occurrence of hydrogen chloride in any stream will increase the corrosion rate, although the relation between the hydrogen chloride concentration in gas streams and corrosion rate is not always linear (Lee and Castaldi, 2008)—the presence of water plays a role. In addition, the effects of hydrogen chloride are promoted by an increase in metal temperature.

5.3 PIPELINES

Pipelines are used to transport crude oil from the wellhead to the refinery. Although many of the difficulties of such transportation are related to the paraffin crystals and construction gel corresponding to the structure of the crude oil rheological features, pipeline corrosion is a more difficult phenomenon that happens due to the instantaneous actions of compositional behavior and properties of the crude oil (Vinay et al., 2005; Yepez, 2005; Ayello et al., 2011; Biomorgi et al., 2012).

A buried pipeline, even one of a relatively short length, will almost inevitably encounter soils that have varying compositions. There can be variations of a physical nature (e.g., differences in coarseness and grain size) as well as variations in type (e.g., rock, loam, and clay). Additional variations can be of a chemical nature, such as pH and chemical constituents. When a pipeline traverses dissimilar soils, the pipeline steel in a particular soil electrolyte will often assume a galvanic potential that is somewhat different from the potential of portions of the same pipeline traversing dissimilar soils elsewhere along the pipeline route. Such galvanic potential differences between different areas of a single pipeline can occur on a macroscale (i.e., over many miles in the route of the pipeline) or on a microscale (within inches of each other or even over shorter distances). A pipeline traversing soils that have varying levels of oxygen concentration will be subject to corrosion cell activity where the portion of the pipeline steel in the area of lowest oxygen concentration is anodic to other areas of the pipeline where there is a greater concentration of oxygen. This form of corrosion activity is also referred to as a concentration cell.

Microorganisms existing in a pipeline trench can affect the control of corrosion either directly or indirectly. Anaerobic bacteria, which thrive in the absence of oxygen, are sulfate-reducing organisms that consume hydrogen and cause a loss of polarization at the steel pipe surface. This loss of polarization can make the attainment of successful cathodic protection much more difficult. Other bacteria that oxidize sulfur can exist in aerated environments. These bacteria (*Thiobacillus thioxidans*) consume oxygen and oxidize sulfides into sulfates, such as sulfuric acid (H_2SO_4). By their metabolic processes, these bacteria can create concentrations of sulfuric acid as high as 10%. Such an environment can be particularly hazardous to pipeline steel.

The electrolytic theory allows an explanation to be made of the corrosion of steel in water (Figure 5.1). The corrosion rate of steel in aerated water doubles for every 30°C (54°F) rise in temperature of the environment in which the steel is confined. However, the effect of pH (acidity or alkalinity) on the corrosion rate is constant from the pH of 10 to the pH of 4, but increases at pH = 3 and becomes catastrophic at pH = 2.5. Also, raising the pH above 10 (increasing the alkalinity) causes the corrosion rate to fall to a minimum. Steel in deaerated and dry soil should not corrode at all and does not when anaerobic bacteria are absent—but most soils are not dry.

The soil resistivity is an indication that moisture and dissolved salts are present, and the corrosivity of the soil is almost proportional to the decrease in resistivity. Soil is considered *noncorrosive* if the resistivity is above 10,000 Ω-cm. Between 1000 and 2000 Ω-cm, it is considered *mildly corrosive*. Between 500 and 1000 Ω-cm, it is in the *corrosive* class. Below 500 Ω-cm is a special situation requiring immediate action, since a typical bare pipeline will corrode in less than a year. This is the *very corrosive* class.

In both offshore and onshore oil field production, the pipelines transporting oil and gas operate with two-phase (oil–water) or three-phase (gas–oil–water) flow conditions (Ajeel and Ahmed, 2008). Crude oil, natural gas, and formation water are transported by risers

A   Direction of Electron Movement in a Pipe

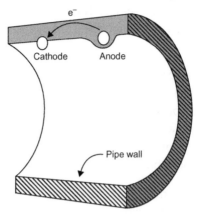

B   Corrosion Mechanisms in a Pipe in the Presence of Water

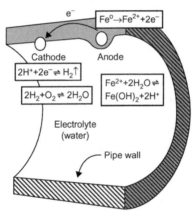

Figure 5.1 Chemical and electrochemical events that contribute to pipeline corrosion: (a) Direction of electron movement in a pipe and (b) corrosion mechanisms in a pipe in the presence of water.

from downhole to wellhead platforms. Well fluids are then carried by flow lines from wellhead platforms to a treatment platform (also called a central gathering platform), where the oil and gas from different wells are mixed. At the treatment platform, the mixtures are separated with separators. Oil and gas are transported separately in pipelines to treatment facilities, storage facilities, and refinery plants. However, even with separation, some produced water—as low as 0.5–2% v/v water cut (volume ratio of water phase to total liquid phase)—is not avoidable in oil transportation pipelines. After separation on the treatment platform, the formation water is injected back into the reservoirs (sometimes together with seawater and carbon dioxide gas) to maintain or increase the reservoir pressure for oil field recovery. In this way, the quantity of produced water in aging oil fields has steadily been growing over the years.

Oil is accompanied by formation water through the whole journey of upstream operations and from reservoirs to downstream facilities. One of the serious problems generated by the formation water is internal corrosion of flow lines and pipelines. Internal corrosion in a pipeline can lead to pipeline failure, loss of production, ultimately shutdown, and a possible environmental disaster.

One of the frequent and major internal corrosion problems is carbon dioxide corrosion (also called sweet corrosion in the oil and gas industry). Carbon dioxide gas dissolves in the produced water and results in the formation of carbonic acid, which gives rise to acid corrosion of carbon steel pipelines. Although corrosion-resistant alloys are produced by pipeline manufacturers to be resistive to sweet corrosion occurring in oil fields, carbon steel is still predominantly used in the design of pipelines and surface facilities in oil fields due to the lower cost of carbon steel versus corrosion-resistant alloys.

The effect of multiphase flow on carbon dioxide corrosion is a complex phenomenon. It is believed that different phase wetting regimes on steel surfaces can greatly affect pipeline internal corrosion. Whenever water physically contacts the pipeline internal surface, which is termed here "water wetting," the steel can corrode. The oil phase plays an important role in internal corrosion control. If the water is fully entrained in the flowing oil and the oil continuously wets the steel surface, where oil wetting occurs, the internal corrosion can be avoided. There is a missing link between different phase wetting regimes and corrosion prediction. Hence, it is a challenge for corrosion engineers to determine more precisely the occurrence of water wetting in multiphase flow leading to carbon dioxide corrosion and conversely oil wetting leading to corrosion-free conditions.

Carbon dioxide corrosion, also called sweet corrosion, involves the interactions between steel, water, and carbon dioxide. In light of common and expensive problems occurring in the oil and gas industry, carbon dioxide corrosion has been extensively studied in the past few decades. Carbon dioxide corrosion is also referred to as *acid corrosion* due to the formation of weak carbonic acid and the release of the hydronium ion (H^+) through several chemical reactions:

$$CO_2(g) \leftrightarrow CO_2(aq)$$
$$CO_2(aq) + H_2O(aq) \leftrightarrow H_2CO_3(aq)$$
$$H_2CO_3(aq) \leftrightarrow H^+ + HCO_3^-$$
$$HCO_3^- \rightarrow H^+ + CO_3^{2-}$$

Compared to strong acids—such as hydrochloric acid (HCl) and sulfuric acid (H_2SO_4), which can completely dissociate in water— carbonic acid is considered a weak acid that dissociates only partially in water. However, aqueous carbon dioxide can lead to a higher corrosion rate than strong acid solutions at the same pH.

Organic corrosion inhibitors have been widely and successfully used to maintain or increase the use of carbon steel in oil fields. They inhibit corrosion by forming an adsorbed organic compound film on the steel surface. In the past, it was well known that the adsorbed compounds work by slowing down reactions occurring in the carbon dioxide corrosion process. However, there are other important effects of inhibitors on phase wetting, which are only considered in a qualitative way. The first effect is a decrease in the oil–water interfacial tension, which relates to the flow pattern of the oil–water two-phase flow. The second effect is a change in the wettability of the steel surface and relates to the hydrodynamic interaction between oil, water, the adsorbed inhibitor film, and the steel surface.

The most common method used for the protection of steel in offshore environments is the use of various types of coatings and—for the immersion zone—coatings combined with cathodic protection.

5.4 OFFSHORE CORROSION

Corrosion is severe in marine atmospheres because of the combined effects of the sun and temperature, along with the oxygen, moisture, and salt contained in the air. This kind of corrosion is generally controlled by use of corrosion-resistant metals and nonmetallic materials where applicable and protective coatings elsewhere (Schremp, 1984).

Seawater is the most corrosive of the natural environments that materials have to withstand, but it is much less corrosive than many environments encountered in the industry, such as mineral acids, as are used in various processes (Chapters 3 and 4). This situation has important implications for desalination materials in that there is a wide range of readily available corrosion-resistant alloys that have been developed for the chemical and process industries and that are resistant to seawater.

The offshore environment is highly corrosive and special stresses put additional pressure on the coating systems used for protection of the steel structures (Braestrup et al., 2005; Guo et al., 2005; Palmer and King, 2008). Offshore maintenance is difficult and the cost is exorbitantly high. It is therefore very important to select quality coating systems and to ensure that they are applied correctly and under the right conditions from the beginning of the construction phase.

Floating offshore production systems—such as *floating production, storage, and offloading* facilities, SPAR (originally designated as a *single point articulated riser*, now generically applied cylindrical floating structures), *tensioned leg platforms*, and *deep draft caisson vessels*— present a number of corrosion control challenges. Corrosion control of such structures is necessary because of: (1) the use of high-strength materials on these structures, (2) the need to control weight and reduce corrosion allowance, (3) the contact of the various parts of the structures, (4) a high number of nonwelded connections, and (5) difficulties in inspection and monitoring.

The corrosion control systems for these types of structures should be specified with caution and a high degree of conservatism. The operator must also be aware of in-service inspection requirements and allow appropriate monitoring and inspection protocols to be specified during construction. For proper performance of coating systems and corrosion protection for such extreme conditions, it is necessary to observe the fundamental parameters of coating selection and the critical parameters that are required to achieve the desired protection. These parameters include: (1) the type and condition of the substrate— usually steel, (2) the environment and possible additional stresses, (3) the quality of any coatings employed, and (4) the quality control, such as application of the coating and the inspection protocols.

5.4.1 Offshore Structures

On surface equipment, the simplest solution is to place an insulating barrier over the metal concerned. Offshore installations are often painted with zinc-rich primers to form a barrier against rain, condensation, sea mist, and spray. The zinc primer not only forms a physical barrier but also acts as a sacrificial anode should the barrier be breached.

Offshore structures are also protected in other ways. The zone above the high tide mark, called the splash zone, is constantly in and out of water. The most severe corrosion occurs here. Any protective coating or film is continually eroded by waves, and there is an ample supply of oxygen and water. Common methods of controlling corrosion in this zone include further coating and also increasing metal thickness to compensate for higher metal loss.

The part of the structure in the tidal zone is subjected to less severe corrosion than the splash zone and can benefit from cathodic protection systems at high tide. Cathodic protection works by forcing anodic areas to become cathodes. To achieve this, a reverse current is applied to counteract the corrosion current. The current can be generated by an external source-impressed cathodic protection or by using sacrificial anodes.

The rest of the structure, which is exposed to less severe seawater corrosion, is protected by cathodic protection. However, crustaceans and seaweed attach to the submerged parts: adding weight that may increase stress-related corrosion. This mechanism occurs when the combined effects of crevice, or pitting, corrosion and stress propagate cracks leading to structural failure. However, a covering does restrict oxygen from reaching the metal thereby reducing corrosion due to oxidation.

Corrosion increases with water salinity: up to about 5% w/w of sodium chloride. Above this, the solubility of oxygen in the water decreases, reducing corrosion rates. In fact, when the salt content is above 15%, the rates are lower than with freshwater. When water and acid gases are present, the corrosion rates rise rapidly. Water dissolves carbon dioxide and/or hydrogen sulfide and becomes acidic.

In highly corrosive environments, carbon steel can be protected by corrosion inhibitors during production. Like acid corrosion inhibitors, these adhere to casing and completion strings to form a protective film. Inhibitors can be continuously introduced into a producing well by a capillary tube running on the outside of the tubing as part of the completion design. Other methods include batch treatment—in which the inhibitor is pumped down the tubing regularly—or squeeze treatments—in which the inhibitor is pumped into the formation.

To protect wells and pipelines from external corrosion, cathodic protection is used. In remote areas, sacrificial ground beds may be used for both wells and pipelines. A single ground bed can protect up to 50 miles (80 km) of pipeline. In the Middle East, solar panels have been used to power impressed current cathodic protection systems. Other methods include thermoelectric generators fueled directly from the pipeline. To protect several wells, a central generator and a distribution network may be used.

Under the right conditions, iron sulfide and iron carbonate scales—the corrosion products when hydrogen sulfide and/or carbon dioxide are present—provide protective coatings. The composition of production fluids, however, typically changes during the life of a reservoir, so relying on natural protection may not be wise. Corrosion monitoring, in some form, should always be undertaken.

Many of the materials used in offshore structures are much more expensive than those used in industries that have traditionally handled seawater, such as shipping and power plants. In these industries, carbon steel, cast iron, copper base alloys, and standard grades of stainless steel have been the usual choice for seawater applications. The desalination industry followed traditional practice and—while in many cases this proved satisfactory—in others it did not. Upgrading, therefore, largely involved economic decisions as the better materials were always available, the main problem was to decide what level of cost and performance was acceptable.

5.4.2 Stainless Steel

There is a wide variety of stainless steels that can be used as condenser tubes in multistage flash distillation units (Figure 5.2). These are used for seawater desalination by flashing a portion of the water into steam in multiple stages of what are essentially countercurrent heat exchangers. The unit is organized into a series of stages, each containing a heat exchanger and a collector for condensate. Each stage has a different pressure, corresponding to the boiling point of water at the stage temperatures. The total evaporation in all the stages is up to approximately 15% v/v of the water flowing through the system, depending on the range of temperatures used. With increasing temperature, there are growing difficulties of scale formation and corrosion; 120°C (250°F)

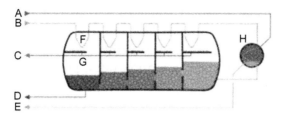

Figure 5.2 Schematic of a multistage flash desalinator.
A: Steam inlet
B: Seawater inlet
C: Potable water outlet
D: Waste outlet
E: Steam outlet
F: Heat exchange area
G: Condensation and collection area
H: Brine heater

appears to be a maximum, although avoidance of scaling on the internals of the unit may require temperatures below 70°C (160°F).

In flowing seawater, the performance of stainless steel tubes is quite satisfactory. They withstand general corrosion and are not affected by the flow velocities operating in a distiller. Normal steels are susceptible to solutions of high chloride content at high temperatures. As the tubes are not subject to any noticeable stresses, this type of attack is unlikely to develop. In stagnant solutions, on the other hand, normal stainless steels are liable to undergo pitting corrosion (Chapter 1). Stainless steel tubes are also prone to crevice corrosion (Chapter 1), which develops under barnacles. Weld areas and the heat-affected zones are weak points where intergranular and pitting attack readily propagates on stainless steel in a desalination unit. Spot and arc welding is to be applied, and annealing should be carried out whenever feasible. Finally, stainless steel tubes mounted on copper base tube plates will lead to their rapid deterioration by galvanic action. Cathodic protection using sacrificial mild steel anodes can be used to overcome this problem.

The part of the structure that is alternately above and below the water line due to tide and waves (usually referred to as the *splash zone*) is subject to even higher corrosion rates: possibly on the order of eight to ten times the corrosion rate of similar structures onshore. Other factors in this area that need consideration include ultraviolet light from the sun, as well as erosion from the water, possible debris, and—in some places of the world—even ice. The part of the structure that is below the water line at lowest tide (usually referred to as the *immersion zone*) is also subject to high corrosion rates: possibly on the order of four to five times the corrosion rate of similar structures onshore.

A special case of immersion is related to the part of the steel structure that is situated below the seabed: typically rammed into the bottom sand. Corrosion rates are much lower due to lower concentrations of oxygen. However, pitting corrosion may also be a factor.

5.4.3 Carbon Steel
Seawater is a complex environment consisting of a mixture of inorganic salts, dissolved gases, and organic compounds. However, it also supports living matter in the form of both macroorganisms (e.g., fish, shellfish, seaweed) and microorganisms. All of these can have an influence on corrosion processes. Seawater is a well-buffered solution, so the pH remains fairly constant—approximately 8 (i.e., slightly alkaline; neutral pH is 7)—and most corrosion processes are dependent on the presence of oxygen.

For carbon steel immersed in seawater, the rate of corrosion is largely dependent on the oxygen content and the temperature of the seawater. The composition of the steel has a relatively minor influence—even when small amounts of alloying elements are added.

5.4.4 Copper Base Alloys
Copper alloys have been traditionally used in marine engineering for heat-exchanger tubing and for cast and wrought components in pumps and valves. These alloys form good protective films in seawater and—provided these films are undamaged—corrosion is slight. However, the protective films are susceptible to damage by fast flowing seawater, and this is an important factor in selecting these alloys for applications involving flow—for example, in heat exchanger tubes.

In deaerated seawater, copper alloys can still form protective films and show higher resistance to corrosion than in natural water because the electrical potential of copper alloys in seawater is much higher than the electrical potential at which hydrogen can evolve: thus, in the absence of oxygen, corrosion is negligible. However, as in the case of carbon steel, low oxygen levels with high flow rates can cause impingement attack, but the rate of attack is much lower than in aerated seawater. The data indicates that at low temperatures and oxygen levels, corrosion of the three alloys is acceptably low. However, at high temperatures with high oxygen levels—although the cupronickels continue to show low corrosion rates—aluminum brass is attacked. Thus, in multistage flash distillation units, where aluminum brass tubing is used, it is confined to the lower temperature recovery stages, with the cupronickels being used in the higher temperatures areas.

Any steel structure placed offshore will be subjected to the stresses of a severely corrosive environment. For example, steel structures situated above the water in the oxygen-containing atmosphere (usually referred to as the *atmospheric zone*) are in a high corrosivity category. The corrosion rate of unprotected steel offshore is high and may be three to four times the corrosion rate of similar structures onshore.

5.4.5 Corrosion by Polluted Water

Polluted seawater contains ammonia and/or hydrogen sulfide. Both materials are products of the decay of dead animals and organisms. The two pollutants attack copper condenser tubes.

Hydrogen sulfide reacts spontaneously with copper tubes to produce a black, porous copper sulfide film. Being nonprotective, the film allows further attack until the metal is eaten through. Attack is more rapid if the polluted water is loaded with slime and silt. Under the slime, attack by sulfide is considered a special type of deposit corrosion in which metal deterioration is initiated by the presence of hydrogen sulfide rather than by a deficiency in oxygen.

Attack by sulfide-polluted water is identified by the black coloration of the inside of the tube. Treatment of the black film with dilute acid releases the hydrogen sulfide, known by its characteristic bad odor. On the microscale range, the presence of sulfide is ascertained through its catalyzing the production of nitrogen gas bubbles from a drop of an iodine–sodium azide solution. The test is carried out under a magnifying lens or microscope.

Sulfide corrosion can be dealt with in a variety of ways. Suitable tube material can be chosen if it is known from the start that only polluted water is available. Extension of the intake facilities away from the source of polluted water is a second alternative. When present in small amounts, hydrogen sulfide is oxidized by chlorination. This must, however, be carefully controlled. The same also applies to rubber ball cleaning if sulfide-polluted sludge is in abundance.

Seawater polluted with ammonia affects the corrosion of copper base condenser tubes in two distinct ways: it induces stainless steel carbon, and the rate of crack propagation increases with pollutant content; and ammonia and ammonium salts enhance the general attack by dissolving and complexing with the copper ions. The tubes lose their protective film and acquire a shiny appearance. Eventually the tubes fail when they become perforated at weak points. Failure can occur anywhere along the tube length and the resulting pore takes any form. Ammonia attack is suspected when daily analysis of the brine shows a constant, high level of copper. There is no adequate solution to the problem of corrosion by ammonia-polluted seawater. If the source of pollution is permanent, relocation of the seawater intake might prove practical.

If pollution is widespread, a change to ammonia-resistant tubes (e.g., titanium or stainless steels) might be considered (Mountford, 2002). Chlorination of ammoniated seawater is effective in removing low levels of pollution. The two agents react to produce chloramines, which are weaker disinfectants than chlorine itself. The formation of chloramines reduces—but does not completely eliminate—the problem of ammonia corrosion, since chloramines readily hydrolyze to the original harmful material. Ferrous sulfate dosing retards the aggressive action of small amounts of ammonia by forming an inhibiting ferric oxide film.

REFERENCES

Ajeel, S.A, Abdullatef Ahmed, M.A., 2008. Study of the synergy effect on erosion-corrosion in oil pipes. Eng. Tech. 26 (9), 1068.

Asteman, H., Spiegel, M., 2007. Investigation of the HCl (g) attack on pre-oxidized pure Fe, Cr, Ni and commercial 304 steel at 400°C. Corrosion Sci. 49 (9), 3626–3637.

Ayello, F., Robbins, W., Richter, S., Nešić, S., 2011. Crude oil chemistry effects on inhibition of corrosion and phase wetting. Paper No. 11060. In: Proceedings of the Corrosion 2011. NACE International, Houston, TX.

Biomorgi, J., Hernández, S., Marín, J., Rodriguez, E., Lara, M., Viloria, A., 2012. Internal corrosion studies in hydrocarbon production pipelines located at Venezuelan Northeastern. Chem. Eng. Res. Des. 90, 1159–1167.

Braestrup, J.B., Andersen, J.B., Andersen, L.W., Bryndum, L.B., Christensen, J.C., Nielsen, N.J.R., 2005. Design and Installation of Marine Pipelines. Blackwell Science Ltd., Oxford, UK.

Garverick, L., 1994. Corrosion in the Petrochemicals Industry. ASM International, Materials Park, OH.

Grabke, H.J., Reese, E., Spiegel, M., 1995. The effects of chlorides, hydrogen chloride, and sulfur dioxide in the oxidation of steels below deposits. Corrosion Sci. 37 (7), 1023–1043.

Guo, B., Song, S., Chacko, J., Ghalambor, A., 2005. Offshore Pipelines. Gulf Professional Publishing, Elsevier, Burlington, MA.

Lee, S.H., Castaldi, M.J., 2008.The effects of varied hydrogen chloride gas concentrations on corrosion rates of commercial tube alloys under simulated environment of WTE facilities. In: Proceedings of the NAWTEC16. 16th Annual North American Waste-to-Energy Conference. 19–21 May, Philadelphia, PA..

Mountford Jr., J.A., 2002. Titanium—properties, advantages and applications solving the corrosion problems in marine service. Paper 02170. In: Proceedings of the Corrosion 2001. NACE International, Houston, TX.

Palmer, A.C., King, R.A., 2008. Subsea Pipeline Engineering, second ed. PennWell Corporation, Tulsa, OK.

Schremp, F.W., 1984. Corrosion prevention for offshore platforms. Soc. Pet. Eng. J. 36 (4), 605–612.

Stott, F.H., Shih, C.Y., 2000. The influence of HCl on the oxidation of iron at elevated temperatures. Mater. Corrosion 51 (5), 277–286.

Vinay, G., Wachs, A., Agassant, J.F., 2005. Numerical simulation of non-isothermal viscoplastic waxy crude oil flow. J. Non-Newtonian Fluid Mech. 128, 144–162.

White, R.A., Ehmke, E.F., 1991. Materials selection for refineries and associated facilities. In: Proceedings of the Corrosion 91. NACE International, Houston, TX.

Yepez, O., 2005. Influence of different sulfur compounds on corrosion due to naphthenic acid. Fuel 84, 97–104.

CHAPTER 6

Corrosion Monitoring and Control

6.1 INTRODUCTION

As refineries produce and process ever more corrosive or erosive hydrocarbon streams, the demands on plant metallurgy steadily increase. Permanently installed sensor systems deliver a continuous update of reactor (or other asset) conditions over time, which can then be correlated with process conditions (Sastri, 1998; Knag, 2005). With such knowledge, refinery operations can move beyond the occurrence of corrosion or erosion and at what rate corrosion is occurring.

Furthermore, analytical testing of process streams is vital to processing high acid crude oils. The monitoring of total acid numbers (TANs) and other relevant properties is of high importance. Acid numbers should be routinely measured in different process streams to optimize dosing of inhibitors (Sastri, 1998; Knag, 2005). Tests involving potentiometric titration are normally used for TAN measurement. Elements—such as trace metals—should be monitored with inductively coupled plasma (ICP) mass spectrometry or ICP optical emission spectrometry instruments. These machines use ICP for elemental analysis.

Some metals in process streams are measured from a corrosion viewpoint. Vanadium (V) and nickel (Ni) are measured to monitor the content of the heavy metals in vacuum gas oil streams that go to the fluid catalytic cracker or to the hydrocracking unit, as these are poisonous to the catalysts in both units. Sometimes they also impact product yields. Additionally, phosphorus content in process streams must be monitored—many inhibitors that are used to control the behavior of high acid crudes are phosphorus-containing compounds. There are restrictions on the allowable limit of phosphorus because of, for example, adverse effects on the catalyst in fluid catalytic cracking units or hydrocracking units.

6.2 CORROSION MONITORING

Combating or preventing corrosion is typically achieved by a complex system of monitoring, preventative repairs, and careful use of materials

(Garverick, 1994). Monitoring methods include both off-line checks taken during maintenance and online monitoring. Off-line checks measure corrosion after it has occurred and when equipment must be replaced based on the historical information that has been collected (*preventative management*). In fact, corrosion monitoring is just as important as recognizing the problem and applying controls. Monitoring attempts to assess the useful life of equipment when corrosion conditions change and how effective the controls are. Techniques used for monitoring depend on what the equipment is, what it is used for, and where it is located.

There are various standard test methods that can be used for the periodic assessment of pipe and vessel integrity. Periodic inspections do not, however, deliver continuous pipework condition data that can be correlated with either corrosion drivers or inhibitors to understand the impact of process decisions and the inhibitor usage on plant integrity. Manual acquisition of ultrasonic wall thickness data is also frequently associated with repeatability limitations and data-logging errors.

Most importantly, seasonal or monthly variations in atmospheric conditions can lead to variations in the environment and, hence, variations in the degree of corrosion (Chapter 1). In locations in which sulfur-containing fuel is not burned during the winter months, this can lead to confusion when test methods are performed during months which are not the climatic norm—although the data will still prove valuable, but conclusions drawn from the data should take into account climatic conditions and climatic changes.

Permanently installed sensor systems, on the other hand, deliver continuous, reliable data. The ultrasonic sensors can be installed on pipes and vessels, operating at up to 600°C (1110°F)—such sensors have also been certified as safe for use in most hazardous environments. Continuous monitoring through use of appropriate test method data can validate that when corrosion is occurring it may be an intermittent process rather than a continuous event. Test methods include the iron powder test method (Hau et al., 2003), which is used for detecting anomalous cases in which oil samples having high acid numbers exhibit less corrosivity than others having much lower acid numbers or in which they show completely different corrosivity despite having similar or the same acid number. In such cases, it is particularly valuable to be able to correlate the data over time with process and/or inhibitor parameters, including fluid dynamics (Cross, 2013).

6.2.1 Corrosion Coupons

The corrosion coupon (*weight loss*) technique is the best known and simplest of all corrosion monitoring techniques. The method involves exposing a specimen of material (the coupon) to a process environment for a given duration, then removing the specimen for analysis. The basic measurement that is determined from corrosion coupons is weight loss; the weight loss taking place over the period of exposure being expressed as corrosion rate. Thus, most data for corrosion of alloys in high-temperature gases are reported in terms of weight change/area for relatively short exposures and inadequately defined exposure conditions. However, metal loss needs to be directly related to the loss of metal thickness used in equipment design and operation decision making. Corrosion in high-temperature gases is affected by parameters of the corrosive environments, such as temperature, alloy composition, time, and gas composition.

Corrosion monitoring is normally done using coupons, electric resistance probes, or linear polarization resistance-based probes that are installed in the overhead line, downstream of the air fin cooler or condensers. Chloride (Cl), pH, and iron (Fe) content in the prefractionator and main fractionator overhead receiver are maintained within limits, and corrosion rates are reduced (Lack and Harrell, 2013). Intermittent wash water dosing in the overhead to remove salts and corrosive substances is also performed. Many refiners use online analyzers for continuous monitoring of pH, iron, and chloride. Continuous measurement presents a step change in the level of corrosion rates that can be determined and in the accuracy of that determination.

The simplicity of the measurement offered by the corrosion coupon is such that the coupon technique forms the baseline method of measurement in many corrosion monitoring programs. The technique is extremely versatile, since weight loss coupons can be fabricated from any commercially available alloy. Also, using appropriate geometric designs, a wide variety of corrosion phenomena may be studied, including, but is not limited to: (1) stress-assisted corrosion, (2) bimetallic (galvanic) attack, (3) differential aeration, and (4) heat-affected zones.

Advantages of weight loss coupons are: (1) that the technique is applicable to all environments: gases, liquids, and solids/particulate flows, (2) that visual inspection can be undertaken, (3) that corrosion deposits can be observed and analyzed, (4) that weight loss can be readily determined and corrosion rate can be easily calculated, (5) that localized corrosion can be identified and measured, and (6) that inhibitor performance can be easily assessed.

Online corrosion coupons have been widely used in critical process circuits for assessment of fluid corrosivity. Normally, alloy-steel and carbon-steel coupons are used for corrosion monitoring and coupons can be fixed or retractable—in retractable coupons at high-temperature locations, a proper sealing system should be used. Coupons can also be of different shapes and types—rectangular, circular, or helical—and may be normal or prestressed. Based on the corrosion rate of coupons (for both general and pitting corrosion), the corrosive nature of the fluid can be assessed and the dosing of additives can be optimized.

In a typical monitoring program, coupons are exposed for a 90-day duration before being removed for a laboratory analysis. This gives basic corrosion-rate measurements at a frequency of four times per year. The weight loss resulting from any single coupon exposure yields the "average" value of corrosion occurring during that exposure. The disadvantage of the coupon technique is that if a corrosion upset occurs during the period of exposure, the coupon alone will not be able to identify the time of occurrence of the upset, and depending upon the peak value of the upset and its duration, may not even register a statistically significant increased weight loss.

Coupon monitoring is most useful in environments where corrosion rates do not significantly change over long time periods. However, they can provide a useful correlation with other techniques, such as electrical resistance and linear polarization resistance methods.

6.2.2 Electrical Resistance Methods

Electrical resistance monitoring involves the use of electrical resistance probes, which can be thought of as *electronic corrosion coupons*. Like coupons, electrical resistance probes provide a basic measurement of metal loss, but unlike coupons, the value of metal loss can be measured at any time, as frequently as required, while the probe is in situ and permanently exposed to the process stream.

The electrical resistance technique measures the change in the resistance (ohms) of a corroding metal element exposed to the process stream. The action of corrosion on the surface of the element produces a decrease in cross-sectional area with a corresponding increase in electrical resistance. The increase in resistance is related to metal loss; and metal loss as a function of time is the corrosion rate. Although still a time averaged technique, the response time for electrical resistance monitoring is far shorter than that for weight loss coupons. Electrical resistance probes have all of the advantages of coupons, plus: (1) direct corrosion rates can be obtained, (2) the probe remains installed in-line until operational life has been exhausted, and (3) they respond quickly to corrosion upsets and can be used to trigger an alarm.

During high acid crude processing, circuits with process fluid temperatures above 180°C (355°F) are the most susceptible to corrosion and, therefore, electrical resistance probes are inserted into the lines. In such cases, the probes are installed at furnace inlet lines of atmospheric and vacuum units, atmospheric gas oil and vacuum gas oil circuits, and reduced crude oils and short residue circuits. They are also used in the feed line to the fluid catalytic cracking unit (preferably in the filter loop) and the hydrocarbon feed line (at approximately 200°C, 390°F) before the charge enters the reactors. More generally, probe locations are identified based on the flow conditions and line geometry. Ideal locations are individual pump discharges, control valve loops downstream of control valves, and in or near thermowells.

Two forms of probe element are available: (1) flush and (2) cylindrical. Also, there are several mounting configurations to choose from, the most common of which allows the probes to be inserted and removed under full process operating conditions without shutdown.

Flush probes are used for the best thermal performance where flush mounting with the pipe wall is desirable or essential. A typical example is a bottom-of-line location. In these applications, water films commonly collect in the bottom of the line and are the primary cause of corrosion. The flush probe ensures that the whole of the probe element is exposed to the water film. Therefore, it is well suited for crude distillation unit (CDU) and vacuum distillation unit (VDU) overhead lines.

Cylindrical probes are suited to virtually any aggressive environment, since there is no sealing material other than the parent metal. The measurement area of the element is much greater in this design and is suitable for use in a single-phase flow.

6.2.3 Field Signature Methods

In field signature methods, corrosion measurements are done directly on the pipe and fittings, whereas electrical resistance probes measure the corrosivity of the process fluid. Field signature methods consist of multiple probes (sensors, metallic pins) that are spot-welded in a rectangular pattern on the pipe/bends at the most critical locations—such as furnace outlets and transfer lines—for monitoring the condition of equipment and piping at high temperatures and fluid velocities. These are normally used where two-phase flow exists at relatively higher temperatures and inaccessible locations.

In the method, an electric current is passed from one side of the matrix to the other, and the voltage between the pins and the critical area is measured. This gives a unique electric field signature that depends on the geometry and thickness of the pipe and the electrical conductivity of the metal. Any change in field (or signature) resulting from internal corrosion or erosion is revealed by a change in voltage across the sensor pins. Online logs are available with permanently installed data-loggers that collect readings on a real-time basis. They have a high level of accuracy, and sensitivity is typically on the order of 0.05–0.1% of pipe wall thickness. The system can be set up to communicate with the refinery control system so data can be reviewed remotely.

A unique advantage of measurement by field signature method is that it gives area coverage, rather than point measurement, and that it detects thickness loss even in the locations between the pins. Other advantages of the method are that the corrosion rate can be accurately measured at high temperatures from remote sections of the refinery or the gas processing plant.

6.2.4 Linear Polarization Resistance Monitoring

The linear polarization resistance monitoring technique is based on complex electrochemical theory. For purposes of refinery applications, the method is simplified to a very basic concept—a small voltage (or polarization potential) is applied to an electrode in solution.

The current needed to maintain a specific voltage shift (typically 10 mV) is directly related to the corrosion on the surface of the electrode in the solution, and the corrosion rate can also be calculated.

The advantage of the linear polarization resistance monitoring technique is that the measurement of corrosion rate is made instantaneously. This is a more powerful than either the coupon method or the electrical resistance method where the fundamental measurement is metal loss and where a measured period of exposure is required to determine corrosion rate. The disadvantage of the method is that it can only be successfully performed in relatively clean aqueous electrolytic environments.

The linear polarization resistance monitoring method is not suitable for gases or water/oil emulsions where fouling of the electrodes will prevent collection of reliable (meaningful) data.

6.2.5 Galvanic Monitoring
The galvanic monitoring technique (*zero resistance ammetry*) is another electrochemical measuring technique. With zero resistance ammetry probes, two electrodes of dissimilar metals are exposed to the process fluid. When immersed in solution, a natural voltage (potential) difference exists between the electrodes. The current generated due to this potential difference relates to the rate of corrosion which is occurring on the more active of the electrode couple. Galvanic monitoring is applicable to the following electrode couples: (1) bimetallic corrosion, (2) crevice and pitting attack, (3) corrosion-assisted cracking, (4) corrosion by highly oxidizing species, and (5) weld decay.

Galvanic current measurement has found its widest application in water injection systems in which dissolved oxygen concentrations are a primary concern. Oxygen leaking into such systems greatly increases galvanic current and, thus, the corrosion rate of steel process components. Furthermore, galvanic monitoring systems are used to provide an indication that oxygen may be invading injection waters through leaking gaskets or deaeration systems.

6.2.6 Other Methods
In addition to field signature methods, several other nonintrusive, online monitoring techniques have become popular in recent years. These alternative systems are often used to validate electrical resistance

probe readings over a longer duration. However, most of these techniques are ultrasonic-based and, therefore, are point contact–type devices. Such techniques do not cover the entire surface area on which the sensors are positioned. These instruments are generally used at locations other than furnace outlets and transfer lines, such as control valve loops and bends at pump discharge (usually on 6- to 8-in diameter lines) and the overhead systems of distillation units.

The unique advantage of such systems is wireless communication. Alternatively, some technologies use permanently installed ultrasonic probes on pipes and equipment to directly measure thickness. These instruments can be clamped onto the pipes, and online data can be gathered. The main disadvantage of these systems is that they measure point thickness and do not capture the area. Also, they are not as accurate or as sensitive as field signature methods. However, these systems can be used in place of electrical resistance probes, and they are less expensive than field signature methods.

6.2.6.1 Iron Powder Test

The *iron powder test* is a method of measuring naphthenic acid corrosion potential and produces results that agree with existing knowledge about this phenomenon (Hau et al., 2003). That higher acid content corresponds to greater corrosivity has been observed in the field and in laboratory test results. Advantageously, the iron powder test has revealed that crude oil samples that have similar or the same or TANs do not show the same corrosivity. The method has also shown that crude oil samples having higher acid number values can exhibit less corrosivity than other crudes having lower acid numbers.

This is because unlike potassium hydroxide (KOH), which does not only react with naphthenic acids but also with other compounds such as hydrolyzable salts, iron naphthenates, inhibitors, and detergents, the iron powder is likely to react more with all of those species also capable of producing corrosion on actual steels. If the naphthenic acids present are stronger, they are expected to produce a larger amount of dissolved iron than in another oil sample having weaker organic acids.

Unlike conventional corrosivity tests based on weight lost and steel coupons, if some corrosion reactions occur that produce insoluble corrosion products (like sulfidation does), the iron powder test should not record this contribution. In this way, the method can give a much better indication of the naphthenic acid corrosion potential than conventional corrosivity tests, since these produce a total corrosion rate that have both the contribution of sulfidation and naphthenic acid corrosion.

6.2.6.2 Hydrogen Penetration Monitoring

In acidic process environments, hydrogen is a by-product of the corrosion reaction. Hydrogen generated in such a reaction can be absorbed by steel, particularly when traces of sulfide or cyanide are present. This may lead to hydrogen-induced failure by one or more of several mechanisms. The concept of hydrogen probes is to detect the amount of hydrogen permeating through the steel by mechanical or electrochemical measurement and to use this as a qualitative indication of corrosion rate. The basis of the technology is that certain modes of corrosion in a refinery result in the generation of atomic hydrogen, and, when atomic hydrogen diffuses through the metal wall and permeates outside the pipe wall, this forms a molecule. By monitoring the changes in the hydrogen diffusion rate, the variation of internal corrosion can be inferred. These portable measuring devices can be used whenever required and can operate at high temperatures.

Monitoring is performed online and off-line. Off-line monitoring refers to the monitoring of the effects of corrosion after it has happened and is usually identified while performing maintenance tasks. While this is not the most effective way of saving the equipment, this was used in older refineries. Currently, an online system has evolved to make it more effective in solving corrosion-related problems.

6.2.6.3 Radiography

Conventional and digital radiography are other proven, nondestructive techniques for assessing piping and equipment health during the processing of high acid crudes. Conventional film radiography can be used as a scanning tool, but generally lacks quantitative capability. However, radiography is not a 24 h per day online monitoring system, although it has mobility that allows it to be deployed at many locations.

In digital radiography, the main advantage is a quantitative approach that provides for online, condition-based assessment of risk. Some machines use solid-state X-ray technology in digital radiography, which is fast, portable, and shows the percentage of metal loss. It does not require scaffolds or insulation removal.

6.2.6.4 Biological Monitoring

The term microbiologically influenced corrosion is used to designate corrosion resulting from the presence and activities of microorganisms within biofilms at metal surfaces.

In aquatic environments, microorganisms attach to metals and form biofilms on the surface. This produces an environment at the biofilm/metal interface that is radically different from that of the bulk medium in terms of pH, dissolved oxygen, and organic and inorganic species, and leading to electrochemical reactions that control corrosion rates. Microorganisms can accelerate rates of partial reactions in corrosion processes and can shift the mechanism for corrosion. Microbiologically influenced corrosion has received increased attention by corrosion scientists and engineers in recent years with the development of surface analytical and electrochemical techniques that can quantify the impact of microbes on electrochemical phenomena and can provide details of corrosion mechanisms. Microbiologically influenced corrosion has been documented for metals exposed to freshwater and to seawater.

Biological monitoring and analysis generally seeks to identify the presence of sulfate reducing bacteria, which is a class of anaerobic bacteria that consume sulfate from the process stream and generate sulfuric acid, a corrosive that attacks production plant materials (Stephen, 1986; Costerton et al., 1987; Little et al., 1991; Videla and Herrera, 2005).

6.2.7 Real-Time Monitoring

Traditional methods used to monitor and control corrosion in the overhead condensing system of an atmospheric crude distillation unit may include installation of corrosion monitoring equipment, use of caustic in the crude oil, and a variety of other chemical corrosion control solutions. Some refiners have elected, at great expense, to upgrade the overhead condensing metallurgy and all associated piping.

These traditional approaches provide acceptable corrosion control during the operating time when the unit is functioning normally. However, the traditional approaches may not detect or allow adequate or timely responses to the corrosion occurrences and the resulting damage. Even the best corrosion-control programs may not detect significant problems before the damage is done.

Online systems are a more modern development and have made substantial differences to the way corrosion is detected and mitigated. There are several types of online corrosion monitoring technologies, such as linear polarization resistance, electrochemical noise, and electrical resistance. Online monitoring has generally had slow reporting rates in the past (minutes or hours) and has been limited by process conditions and sources of error, but newer technologies can report rates up to twice per minute with much higher accuracy (referred to as real-time monitoring). This allows process engineers to treat corrosion as another process variable that can be optimized in the system. Immediate responses to process changes allow the control of corrosion mechanisms, so they can be minimized while also maximizing production output (Uhlig and Revie, 1985; NACE, 2002; Kane et al., 2005; Hilton and Scattergood, 2010). In an ideal situation, having online corrosion information that is accurate and real time will allow conditions that cause high corrosion rates to be identified and reduced (*predictive management*).

The effectiveness of traditional testing is limited by the time it takes to collect and analyze samples. These tests are generally part of the routine service performed by refinery operations personnel or the chemical supplier, and they may only be performed at daily or weekly intervals. The frequency of sampling and performing wet chemistry tests for chloride, iron, and ammonia is often less. The result may be a limited number of data sets per year—the majority collected during periods of stable operation when little or no corrosion occurs. The same limitations apply to corrosion-rate data collected from probes and other monitoring devices in the overhead. Test accuracy and speed of results turnaround are significant concerns when relying on manual wet chemistry testing. Human error, choice of test method, and the temptation to take shortcuts in sampling technique and preparation can significantly affect test accuracy. The key to controlling corrosion, without throwing metallurgy at the problem, is the ability to capture accurate data in real time, detecting and closing the corrosion window before significant damage occurs.

Real-time monitoring greatly enhances the ability of the refiner to detect and take timely action to correct variations in process conditions that can lead to corrosion. To be truly effective, a corrosion-control program must go beyond the industry current practice of periodic sampling and manual sample processing. For example, when installed into a crude distillation unit, an overhead stream analyzer can provide continuous, accurate, repeatable data, including conditions during the critical parts of the operation. By detecting *corrosion windows* consistently and in real time, the analyzer provides the refiner with a continuous view of pH, chloride, and iron levels in the system, permitting the application of timely and effective chemical solutions before significant corrosion has occurred.

6.3 CORROSION CONTROL AND INHIBITION

Corrosion control is an ongoing, dynamic process in the prevention of metal deterioration in three general ways: (1) changing the environment, (2) changing the material, and (3) placing a barrier between the material and its environment. The material does not have to be metal—but *is* in most cases. The metal does not have to be steel, but, because of the strength, ready availability, and cheapness of this material, it usually an alloy of steel. Again, the environment is, in most cases, the atmosphere, water, or the earth, and is an important contributor to corrosion chemistry.

Many of the methods for preventing or reducing corrosion exist, most of them oriented in one way or another toward slowing rates of corrosion and reducing metal deterioration (Bradford, 1993; Jones, 1996). All methods of corrosion control are variations of these general procedures, and many combine more than one of them. The material does not have to be metal but *is* a metal or an alloy of metals in most cases. There are, however, enough exceptions to make corrosion control more complex than the simple chemical equations would indicate.

6.3.1 Cathodic Protection

Cathodic protection is a technique used to control the corrosion of a metal surface by making it the cathode of an electrochemical cell (Chapter 1) (Peabody, 2001; Bushman, 2002). Cathodic protection can, in some cases, prevent stress corrosion cracking, and the method is used widely to protect buried or submerged pipeline. This chemistry employed in cathodic protection seeks to reduce the rate of corrosion of the structure to be protected by joining it to *sacrificial* anodes in which the structure is joined to another metal (an anode) that corrodes more readily, effectively diverting the tendency to corrode away from the structure.

Cathodic protection uses direct electrical current to counteract the normal external corrosion of a metal pipeline. In the process, a structural metal, such as iron, is protected from corrosion by connecting it to a metal that has a more negative reduction half-cell potential, which now corrodes instead of iron. There are two major variations of the cathodic method of corrosion protection. The first is called the impressed current method, and the other is called the sacrificial anode method.

As an alternative to using metals that must be protected by one or other of the methods described, there is also the option of using an alloy selected for having a greater resistance to corrosion caused by its surroundings. However, alloys with good resistance in one environment may have poor resistance in another, and their resistance is also likely to vary according to differences in exposure conditions, such as temperature or stress.

However, a side effect of improperly applied cathodic protection is the production of hydrogen ions, leading to its absorption in the protected metal and subsequent hydrogen embrittlement of welds and materials with high hardness—such as might occur in pipelines. Under typical conditions, the ionic hydrogen will combine at the metal surface to create hydrogen gas, which cannot penetrate the metal. Hydrogen ions, however, are small enough to pass through the crystalline steel structure, and this can lead in some cases to hydrogen embrittlement.

Cathodic disbonding is the process of disbondment of protective coatings from the protected structure (cathode) due to the formation of hydrogen ions over the surface of the protected material (cathode). Disbonding can be exacerbated by an increase in alkali ions and an increase in cathodic polarization. The degree of disbonding is also reliant on the type of coating, with some coatings being effected more than others. Cathodic protection systems should be operated so that the structure does not become excessively polarized, since this also promotes disbonding due to excessively negative potentials. Cathodic disbonding occurs rapidly in pipelines that contain hot fluids (as might occur in transfer pipes in a refinery) because the process is accelerated by heat flow.

Cathodic protection has been used since the middle of the 19th century and has gained widespread acceptance (Garverick, 1994). Anodic protection is sometimes confused with cathodic protection, but the two techniques are fundamentally different with respect to which electrode is protected: the cathode is protected in cathodic protection and the anode is protected in anodic protection. Anodic protection is used to a lesser degree than the other corrosion control techniques, particularly cathodic protection, because of the limitations on metal–chemical systems for which anodic protection will reduce corrosion. Anodic protection does have a place in the corrosion control area, for which it is a feasible technique (Kuzub and Novitskiy, 1984).

6.3.2 Galvanic Cathodic Protection

In order for galvanic cathodic protection to work, the anode must possess a lower (i.e., a more negative) potential than that of the cathode (the structure to be protected). Thus, galvanic anodes are designed and selected to have a more negative electrochemical potential than the metal of the structure (typically steel). For effective cathodic protection, the potential of the steel surface is polarized (pushed) more negative until the surface has a uniform potential. At that stage, the driving force for the corrosion reaction is removed. The galvanic anode continues to corrode and the anode material is consumed until it must be replaced. The polarization is caused by the electron flow from the anode to the cathode. The driving force for the cathodic protection current is the difference in electrochemical potential between the anode and the cathode (Roberge, 1999).

Galvanic or sacrificial anodes are made in various shapes and sizes using alloys of zinc, magnesium, and aluminum.

6.3.3 Impressed Current Cathodic Protection

For larger structures, galvanic anodes cannot economically deliver enough current to provide complete protection. Impressed current cathodic protection systems use anodes connected to a direct current (DC) power source. Usually this will be a cathodic protection rectifier, which converts an alternating current (AC) power supply to a DC output.

Anodes for impressed current cathodic protection systems are available in a variety of shapes and sizes. Common anodes are tubular and solid rod shapes or continuous ribbons of various materials. These include materials such as high silicon cast iron, graphite, and mixed metal oxides.

6.3.4 Cathodic Shielding

The effectiveness of cathodic protection systems on steel pipelines can be impaired by the use of solid film backed dielectric coatings such as polyethylene tapes, shrinkable pipeline sleeves, and factory applied single or multiple solid film coatings. The protective electric current from the cathodic protection system is blocked or shielded from reaching the underlying metal by the highly resistive film backing.

With the development of this cathodic protection shielding phenomenon, impressed current from the cathodic protection system cannot access exposed metal under the exterior coating to protect the pipe surface from the consequences of an aggressive corrosive environment. The cathodic protection shielding phenomenon induces changes in the potential gradient of the cathodic protection system across the exterior coating, which are further pronounced in areas of insufficient or substandard cathodic protection current emanating from the cathodic protection system of the pipeline. This produces an area on the pipeline of insufficient cathodic protection defense against metal loss aggravated by an exterior corrosive environment.

6.4 CORROSION CONTROL IN REFINERIES

For practical purposes, corrosion in refineries can be classified into low- and high-temperature corrosion. Low-temperature corrosion is considered to occur below approximately 260°C (500°F) in the presence of water. Carbon steel can be used to handle most hydrocarbon streams in this temperature range, except where aqueous corrosion by inorganic contamination, such as hydrogen chloride or hydrogen sulfide, necessitates selective application of more resistant alloys. High-temperature corrosion is considered to take place above approximately 260°C (500°F). The presence of water is not necessary because corrosion occurs by the direct reaction between metal and environment.

The major cause of low-temperature (and, for that matter, high-temperature) refinery corrosion is the presence of contaminants in crude oil as it is produced. Although some contaminants are removed during preliminary treating at the wellhead fields as well as during dewatering and desalting, they can still appear in refinery tankage, along with contaminants picked up in pipelines or marine tankers. However, in most cases, the actual corrosives are formed during initial refinery operations. For example, potentially corrosive hydrogen chloride evolves in crude preheat furnaces from relatively benign calcium chloride ($CaCl_2$) and magnesium chloride ($MgCl_2$) entrained in crude oil (Samuelson, 1954). Mitigation of the problems related to low-temperature corrosion are associated with adequate cleaning of the crude oil and the removal of corrosive contaminates at the time of, or immediately after, formation.

The pH stabilization technique can be used for corrosion control in wet gas pipelines when no or very little formation water is transported in the pipeline. This technique is based on precipitation of protective corrosion product films on the steel surface by adding pH-stabilizing agents to increase the pH of the water phase in the pipeline. This technique is very well suited for use in pipelines where glycol is used as a hydrate preventer, as the pH stabilizer will be regenerated together with the glycol—thus, there is very little need for replenishment of the pH stabilizer.

In order to overcome the effects of corrosion, it is necessary to use of specific types of metal to be durable in spite of the effects of corrosion. Carbon steel is used for the majority of refinery equipment requirements as it is cost efficient and withstands most forms of corrosion due to hydrocarbon impurities below a temperature of 205°C (400°F), but as it is not able to resist other chemicals and environments, it is not used universally. Other kinds of metals used are low alloys of steel containing chromium and molybdenum, and stainless steel containing high concentrations of chromium for excessively corrosive environments. More durable metals such as nickel, titanium, and copper alloys are used for the most corrosive areas of the plant that are exposed to the highest of temperatures and the most corrosive of chemicals (Burlov et al., 2013).

Many problems related to the correct usage of corrosion control measures (e.g., injection of chemicals such as inhibitors, neutralizers, biocides, and others) may be solved by means of corrosion monitoring methods (Groysman, 1995, 1996, 1997). For example, hydrocarbons containing water vapors, hydrogen chloride, and hydrogen sulfide leave the atmospheric distillation column at 130°C (265°F). This mixture becomes very corrosive when cooled below the dew point temperature of 100°C (212°F). In order to prevent high acidic corrosion in the air cooler and condensers, neutralizers and corrosion inhibitor are injected in the overhead of the distillation column. In addition, corrosion monitoring equipment such as weight loss coupons and electrical resistance probes should be installed in several places, including the naphtha pump-around and kerosene pump-around lines. The electrical resistance probes show the corrosion situation continuously. The weight loss should be changed every several months, in order to compare with the results of the electrical resistance probes and to examine the danger of chloride attack (pitting corrosion). The more points in the unit used for corrosion monitoring, the better and more efficient is the corrosion coverage.

In summary, control of corrosion requires: (1) evaluation of the potential corrosion risks, (2) consideration of control options—principally inhibition as well as materials selection, (3) monitoring whole life cycle suitability, (4) life cycle costing to demonstrate economic choice, and (5) diligent quality assurance at all stages.

6.4.1 Crude Oil Quality

Crude oil quality is an important aspect of corrosion that is often not recognized as often as other causes (Chapter 3). Crude oil value—to a refinery—is based on the expected yield and value of the products value, less the operating costs expected to be incurred to achieve the desired yield. Ensuring that the quality of crude oil received is equivalent to the purchased quality (value acquired is equal to value expected) is—with the growing popularity of heavy feedstocks, opportunity crudes, and high acid crudes—one of the greatest challenges facing the refining industry (Cross, 2013; Vetters and Clarida, 2013).

Furthermore, difficulties in minimizing differences between purchased quality and refinery-receipt quality are significantly higher when multiple crude oils are processed as a blend, and the complexity of the crude delivery system increases. Shipping crude oil through multiple pipelines and redistribution storage tanks—a reality faced by most inland refiners—results in the delivered crude oil being a composite of the many crude oils. Thus, the resultant composite blend may vary significantly from the expected purchased quality, and the sources of quality problems are difficult to estimate.

Issues regarding crude oil properties occur regardless of whether the dominant crude slate is comprised of domestic crude delivered by pipeline or foreign crude delivered via waterborne transportation. In both cases, using simple categories such as gravity or sulfur do not provide an accurate measure of the particular crude oil value, and monitoring only gravity and sulfur does not provide adequate safeguards for the integrity of the crude oil while it is in transit. More sophisticated analyses (with an analysis for constituents likely to cause corrosion and correlated to refinery performance) can provide a comprehensive estimate of quality value to a specific refinery. This analysis needs to give weight to quality consistency, where appropriate, as well as to improved yield, reduced operating expenses, and the compatibility of the crude oil feedstock to the refinery processing hardware.

In addition, a combination of aging plants, greater fluid corrosiveness, and the tightening of health, safety, security, and environmental requirements has made corrosion management a key consideration for refinery operators. The prevention of corrosion erosion through live monitoring provides a real-time picture of how the refinery is coping with the high demands placed upon it by corrosive fluids. This information can assist in risk management assessments.

6.4.2 Acidic Corrosion

Refinery equipment reliability during the processing of high acid crude oils is paramount. Hardware changes—such as upgrading materials construction from carbon steel (CS) and alloy steel to stainless steel (SS) 316/317, which contains molybdenum and is significantly resistant to naphthenic acid corrosion—are complicated tasks and require large capital investment as well as a long turnaround for execution. Alternatives to hardware changes are corrosion mitigation with additives and corrosion monitoring with the application of inspection technologies and analytical tests.

During naphthenic crude processing, corrosion at high temperatures is mitigated by injecting either phosphate-based ester additives or sulfur-based additives, which provide an adherent layer that does not corrode or erode due to the effect of naphthenic acids. It has been suggested that corrosion during processing of high acid crude oils is a lower risk of the sulfur content is high—the relationship between the acid number and amount of sulfur is not fully understood, but it does appear that the presence of sulfur-containing constituents has an inhibitive effect (Piehl, 1960; Mottram and Hathaway, 1971; Slavcheva et al., 1998, 1999).

Mitigation of process corrosion includes blending, inhibition, materials upgrading, and process control. Blending may be used to reduce the naphthenic acid content of the feed, thereby reducing corrosion to an acceptable level. Blending of heavy and light crudes can change shear stress parameters and might also help reduce corrosion. Blending is also used to decrease the level of sulfur content in the feed and to inhibit, to some degree, naphthenic acid corrosion (Chapter 1).

6.4.3 Sulfidic Corrosion

Any of the 18Cr−8Ni stainless steel grades can be used to control sulfidation. Some sensitization is unavoidable if exposure in the sensitizing temperature range is continuous or long term. Stainless steel equipment subjected to such exposure and to sulfidation corrosion should be treated with a 2% w/w soda ash solution or an ammonia solution immediately upon shutdown to avoid the formation of polythionic acid, which can cause severe intergranular corrosion and stress cracking.

Vessels for high-pressure hydrotreating and other heavy crude fraction upgrading processes (e.g., hydrocracking) are usually constructed of one of the Cr−Mo alloys. To control sulfidation, they are internally clad with one of the 300 series stainless steels by roll or explosion bonding or by weld overlay. In contrast, piping, exchangers, and valves exposed to high-temperature hydrogen−hydrogen sulfide environments are usually constructed of solid 300 series stainless steel alloys. In some designs, Alloy 800H has been used for piping and headers. In others, centrifugally cast HF-modified piping has been used. High nickel alloys are rarely used in refinery or petrochemical plants in hydrogen−hydrogen sulfide environments because of their susceptibility to the formation of deleterious nickel sulfide. They are particularly susceptible to this problem in reducing environments. As a general rule, it is recognized that the higher the percentage of nickel in the alloy, the more susceptible the material to corrosion.

Vapor diffusion aluminum coatings (*alonizing*) have been used with carbon, Cr−Mo, and stainless steels to help control sulfidation and reduce scaling. For the most part, this has been restricted to smaller components. Aluminum metal spray coatings have also been used, but not widely nor very successfully.

6.4.4 Carburization

Carburization is not a common occurrence in most refining operations because of the relatively low tube temperatures of most refinery fired heaters. However, it can and does occur in those higher temperature process heaters (e.g., cokers) where its control in 9Cr−Mo tubes has been somewhat successful through the use of an aluminum vapor diffusion coating. Carburization can also occur during an upset that results in the exposed material being heated to unacceptably high temperatures.

When it is expected to occur within the range of normal operating conditions, the broader approach has been to use Type 304H for temperatures up to about 815°C (1500°F). There is no advantage in using either of the stabilized grades since any unreacted titanium or niobium from the original melt would be quickly tied up. Type 310 or Alloy 800H may be used for temperatures up to about 1010°C (1850°F). For the most part, refinery application of the latter two alloys for this purpose is confined to hydrogen reformer furnaces. Unfortunately, the 300 series stainless alloys, including Type 310, are subject to sigma phase embrittlement in the temperature range in which they have useful carburization resistance. Alloy 800H is a better choice.

The most effective element in controlling carburization is nickel in combination with chromium. Silicon also has a strong effect; and aluminum in excess of 3.5−4% w/w is also beneficial. Unfortunately, the presence of much more than 2% w/w silicon adversely affects the rupture strength and weldability of both wrought and cast heat-resistant alloys. Aluminum, in concentrations greater than 2−2.5% w/w, has an adverse effect on ductility and fabrication properties that are essential for piping, tubing, and pressure vessels.

Coatings and surface enrichment using silicon, aluminum, chromium, and combinations thereof have been tried to control carburization of heat-resistant alloys. Unfortunately, none of these have been successful in the long term. Vapor-diffused aluminum enrichment has shown promise and performed well at lower temperatures, but it breaks down after relatively short times at temperatures above 1010−1040°C (1850−1905°F).

Carburization is far more common in the petrochemical industry than in refining. The most common occurrence is in the radiant and shield sections of ethylene cracking furnaces. Carburization is a serious problem in these furnaces because of the high tube metal temperatures—up to 1150°C (2100°F)—and the high carbon potential associated with the ethane, propane, naphtha, and other hydrocarbon feedstocks that are cracked. However, it also occurs, albeit less frequently and with less severity, in reforming operations and in other processes handling hydrocarbon streams or certain ratios of $CO/CO_2/H_2$ gas mixtures at high temperatures.

The form and severity of decoking operations appear to play important roles in the rate of carburization. High-temperature decoking with low quantities of steam are thought to accelerate carburization. Likewise, steam—air decoking appears to be more deleterious than steam only. Appropriate metallurgy can be used to reduce carburization and the important characteristic of an alloy is its ability to form a stable, protective oxide film. Chromium oxide is considered to be such a film. However, it is not sufficiently stable at higher operating temperatures and low oxygen partial pressures. Alumina or silica are much better suited for those conditions. Unfortunately, the addition of aluminum or silicon to the heat-resistant alloys in quantities to develop full protection involves trade-offs in strength, aged ductility, and/or weldability that are often unacceptable. Viable alloys are generally restricted to about 2% of either element.

6.4.5 High-Temperature Corrosion

High-temperature crude corrosion is a complex problem. There are at least three corrosion mechanisms: (1) furnace tubes and transfer lines, where corrosion is dependent on velocity and vaporization and is accelerated by naphthenic acid, (2) vacuum columns, where corrosion occurs at the condensing temperature, is independent of velocity, and increases with naphthenic acid concentration, and (3) side-cut piping, where corrosion is dependent on naphthenic acid content and is inhibited somewhat by sulfur compounds (Chapter 1).

To mitigate *high-temperature corrosion*, a high-temperature corrosion inhibitor is dosed into the process streams, normally at a concentration of 3–15 ppm w/w. The additive is mixed with the suitable process stream, typically at a ratio of 1:30, and the mix is injected into the cooled stream ($<100°C$, $<212°F$) with specially designed corrosion-mitigation quills, since the inhibitor is highly corrosive to alloy steel and even stainless steel. The dosage limit of the phosphorus-based inhibitor is calculated from the allowable phosphorus level in the products—products such as the vacuum gas oil that is used as the feedstock for a hydrocracking unit or as the feedstock for a fluid catalytic cracking unit.

In terms of high acid crude oils, corrosion is predominant at temperatures higher than 180°C (355°F), where shear stress on pipe walls is significant. Such corrosion problems in the high-temperature section of the atmospheric distillation column are normally mitigated by dosing phosphate ester-based or sulfur (S)-based inhibitors at certain critical locations inside the process units. The inhibitors are dosed in process streams with the help of injection quills.

The first phase of an engineered solution is to perform a comprehensive high TAN impact assessment of a crude unit processing a target high TAN blend under defined operating conditions. An important part of any solution system is the design and implementation of a comprehensive corrosion monitoring program. Effective corrosion monitoring helps confirm which areas of the unit require a corrosion-mitigation strategy and provides essential feedback on the impact of any mitigation steps taken.

With a complete understanding of the unit operating conditions, crude oil and distillate properties, unit metallurgies, and equipment performance history, a probability of failure analysis can be performed for those areas that would be susceptible to naphthenic acid corrosion. Each process circuit is assigned a relative failure probability rating based on the survey data and industry experience.

Corrosion inhibitors are often the most economical choice for mitigation of naphthenic acid corrosion. Effective inhibition programs can allow refiners to defer or avoid capital intensive alloy upgrades, especially where high TAN crudes are not processed on a full-time basis. Baker Petrolite has pioneered the use of low phosphorous inhibitor formulations in refineries in which the potential downstream effects of phosphorous are a concern. Use of best practices for high-temperature inhibitor applications ensures that the correct amount of inhibitor is delivered safely and effectively to all of the susceptible areas of the unit.

Injection of corrosion inhibitors may provide protection for specific fractions that are known to be particularly severe. Monitoring in this case needs to be adequate to check on the effectiveness of the treatment. Process control changes may provide adequate corrosion control if there is the possibility of reducing charge rate and temperature. For long-term reliability, upgrading the construction materials is the best solution. Above 290°C (555°F), with very low naphthenic acid content, cladding with chromium (Cr) steels (5−12% w/w Cr) is recommended for crudes of greater than 1% w/w sulfur when no operating experience is available. When hydrogen sulfide is evolved, an alloy containing a minimum of 9% w/w chromium is preferred. In contrast to high-temperature sulfidic corrosion, low alloy steels containing up to 12% w/w Cr do not seem to provide benefits over carbon steel in naphthenic acid service. Type 316 stainless steel (>2.5% w/w molybdenum) or Type 317 stainless steel (>3.5% w/w molybdenum) is often recommended for cladding of vacuum and atmospheric distillation columns.

Alloys intended for high-temperature applications are designed to have the capability of forming protective oxide scales. Alternatively, when the alloy has ultrahigh-temperature strength capabilities (which is usually synonymous with reduced levels of protective scale-forming elements), it must be protected by a specially designed coating. Oxides that effectively meet the criteria for protective scales and can be formed on practical alloys are limited to chromia (Cr_2O_3), alumina (Al_2O_3), and silicon dioxide (SO_2). In the pure state, alumina exhibits the slowest transport rates for metal and oxygen ions and so should provide the best oxidation resistance.

A useful concept in assessing the potential high-temperature oxidation behavior of an alloy is that of the reservoir of scale-forming element contained by the alloy in excess of the minimum level (approximately 20% w/w for iron−chromium alloys at 1000°C (1832°F)). The more likely the service conditions are to cause repeated loss of the protective oxide scale, the greater the reservoir of scale-forming element required in the alloy for continued protection. Extreme cases of this concept result in chromizing or aluminizing to enrich the surface regions of the alloy or in the provision of an external coating rich in the scale-forming elements.

6.4.6 Cooling Water Corrosion
Also, problems exist in every cooling water system in the oil refining industry, including: (1) corrosion, (2) inorganic deposits containing carbonate scale, (3) corrosion products of iron, phosphates, silicates, and

some others, and (4) biofouling (microbial contamination). Online corrosion and deposit monitoring systems used in the cooling water system allow monitoring of the general corrosion of carbon steel (or any other alloy) at ambient temperatures (nonheated steel surface) and at the drop temperature in the heat exchanger (heated steel surface), the pitting tendency for heated and nonheated surfaces, and the heat transfer resistance—the quantitative value of inorganic and organic deposits (fouling).

One of the major contributors to corrosion is the pH value of process water: pH measurements in the oil refinery industry are not always satisfactory due to the poor ability to measure the extent of corrosion in certain environments. When the correct equipment is chosen, however, in-line pH measurement and control facilities have proven to be of great value in reducing plant-wide corrosion and reducing the use of chemicals such as pH control reagents and corrosion inhibitors. This not only results in significant cost savings, but also in increased earnings through increased on-stream process time.

6.5 CORROSION CONTROL IN GAS PROCESSING PLANTS

As in other situations, control of the corrosion rate in gas processing plants can be affected by reducing the tendency of the reactor or pipe metal to oxidize, by reducing the aggressiveness of the medium, or by isolating the metal from the fluid. The latter can be achieved by coating the metal with a thick impervious noncorroding coating. Such coating, provided there is no spurious interaction with the reactor or pipe contents, has wide use, but the effect of the coating may not be permanent because of breaks in the coating over time. In addition, in some systems, the coating might interfere with the process for which the equipment is used by changing the heat transfer properties of the metal.

6.5.1 Steel Corrosion

Mild steel has been the most widely used alloy for structural and industrial applications since the beginning of the industrial revolution. The use of acid media in the study of corrosion of mild steel has become important because of its use in industrial applications such as acid pickling, industrial cleaning, acid descaling, oil-well acid in oil recovery, and petrochemical processes. Crude oil refining is carried out under a variety of corrosive conditions and, in such, the corrosion of equipment is generally caused by a strong acid attacking an equipment surface. And in many of the structural and industrial applications of mild steel, they are also exposed to corrosive environments and they are susceptible to different types of corrosion. Therefore, the use of corrosion inhibitors to prevent metal dissolution will be inevitable. The use of inhibitors is found to be one of the most practical methods for protection against corrosion, especially in acidic media. The majority of well-known inhibitors are organic compounds containing heteroatoms (nitrogen, oxygen, sulfur) and multiple bonds.

In cases in which a fairly thick coating is not acceptable, the use of corrosion inhibitors comes into play. These chemicals are continually fed into the fluid with the objective of having them move to the metal–fluid interface. There the intact inhibitor molecularly attaches to the metal or it reacts with the surface to form a thin adherent compound. In the first case they act by adsorption. In either case, the films are only one to a few molecules thick (i.e., nanometer thickness).

For many aqueous systems, for elevated temperature equipment, for crude oil production, organic (carbon-based) chemicals find considerable use. These materials are likely to function by adsorption. Here the organic molecule, which orients itself suitably, becomes attached to the solid surface—often via a less than total reaction between the inhibitor molecule and the solid surface. The attachment does not require a total electron transfer in either direction. Ion–dipole attraction suffices to attach the inhibitor molecule to the solid surface, which interferes with access of the corrosive entity to the surface. The adsorbed layer can be formed all over the surface either in a single layer, a multilayer, or a mixture of both. This process has the advantage of being molecularly thin and thus not too intrusive, for example, in heat conduction.

However, the amount of a given material adsorbed from a mixture depends on its concentration, temperature, and fluid flow rate, as well as on the nature of the adsorbent, that is, the solid surface. The film has to be kept intact by continually adding inhibitor to the medium to maintain a predetermined concentration of inhibitor. The continuing concentration is generally lower than that used initially, but both are at the millimolar level. Furthermore, if the temperature or flow rate of the system changes, the amount adsorbed is apt to change. For temperature, the change is energetic: a basic change in the amount adsorbed.

For fluid flow, the change is dynamic: that is a change of fluid movement at the interface, which may affect the amount left at the interface. Faster flow generally causes removal of some of the physically adsorbed material from the solid surface.

Corrosion inhibitors are of limited value if they do not reach the metal surface intact. They can be lost by reaction with chemicals in the stream or to any other solid surface exposed to the fluid stream, such as sand. Also, the inhibitor should not be detrimental to either process or product. Further, since the system may change with time, so must the corrosion control, which must be monitored consistently.

6.5.2 Flue Gas Desulfurization

Flue gas desulfurization (FGD) is a very common method for gas claning adopted in a gas processing plant. In this process, the flue gas with acid vapors is scrubbed to remove it as a by-product. Most of the FGD processes use alkali to scrub the flue gas. Many designers of FGD adopt the limestone gypsum process. This process has gained acceptance due to the saleable gypsum by-product. Seawater availability makes it possible for use as an absorbent of sulfur oxides in acid form. There is another process called the Wellman–Lord process, which is a regenerative process that uses an aqueous sodium sulfite solution for scrubbing flue gas. The saleable by-product, depending on the plant's design, could be elemental sulfur, sulfuric acid, or liquid sulfur dioxide. The *sodium bicarbonate injection process* is a direct injection method adapted to desulfurize the flue gas by injection of sodium bicarbonate into the duct after the air preheater and before the dust removal system, like an electrostatic precipitator or bag filters.

Ammonia injection has been adopted as a method of sulfur control in certain process plant boilers burning high sulfur oil due to the availability of ammonia. Ammonia is injected into the economizer region, where the temperature of flue gas is below the ammonia dissociation temperature and sufficient time is available for the chemical reaction. Ammonia combines with sulfur trioxide to form ammonium sulfate. The rate of ammonia injection will depend upon the concentration of sulfur trioxide. The problem with this method is it produces a high volume of loose deposits of ammonium sulfate, which increases the pressure drop in the flue gas path. Removal of these deposits is achieved by water washing the air preheater online.

6.6 CORROSION CONTROL IN PIPELINES

Four common methods used to control corrosion in pipelines are: (1) protective coatings and linings, (2) cathodic protection, (3) materials selection, and (4) inhibitors.

Coatings and linings are the principal tools for defending against corrosion and are often applied in conjunction with cathodic protection systems to provide the most cost-effective protection for pipelines. Cathodic protection is a technology that uses direct electrical current to counteract the normal external corrosion of a metal pipeline and is used where all or part of a pipeline is buried underground or submerged in water. Materials selection refers to the selection and use of corrosion-resistant materials, such as stainless steels, plastics, and special alloys, to enhance the life span of a structure such as a pipeline. Corrosion inhibitors are substances that, when added to a particular environment, decrease the rate of attack of that environment on a material such as metal or steel-reinforced concrete. Corrosion inhibitors can extend the life of pipelines, prevent system shutdowns and failures, and preclude product contamination. Evaluating the environment in which a pipeline is or will be located is very important to corrosion control, no matter which method or combination of methods is used. Modifying the environment immediately surrounding a pipeline, such as reducing moisture or improving drainage, can be a simple and effective way to reduce the potential for corrosion.

The selection of materials for pipeline construction is limited when all of the aspects of safety, structural integrity, operating life, and economic considerations are taken into account and acted upon. Carbon steel is the almost exclusive choice of pipeline designers. This is true for pipeline systems that are used to gather or collect the natural gas, crude oil, or water; it is also true for those pipelines that are used to transport substances over distances of hundreds of feet to hundreds of miles. It is also the case for piping systems that are used to distribute natural gas, water, water-refined liquids, and so on, to the end user.

Cast iron is extensively used in water and natural gas distribution systems. In recent years, nonmetallic materials have found application in natural gas distribution systems as carrier vehicles and as liners for restoring failed metallic piping to service without the need for trenching and replacement. Carbon steels and, in certain types of service or environmental conditions, alloy steels are by far the most commonly used pipeline materials of construction.

6.6.1 Protective Coatings

The function and desired characteristics of a protective (dielectric-type) pipeline coating is to control corrosion by isolating the external surface of the underground or submerged piping from the environment, to reduce cathodic protection requirements, and to improve (protective) current distribution. Coatings must be properly selected and applied, and the coated piping must be carefully installed to fulfill these functions. Different types of coatings can accomplish the desired functions. The desired characteristics of the coatings are: (1) effective electrical insulation—to prevent the electrolytic discharge of current from the steel surface of the pipe, the coating must have the characteristics of an effective dielectric material, (2) effective moisture barrier—the permeation of a coating material by soil moisture would significantly reduce its dielectric properties, and (3) application considerations—the coating must be capable of being readily applied to the pipe, and it must not of itself or through required application procedures adversely affect the properties of the pipe. Pipe coated with a galvanic coating, such as zinc (galvanizing) or cadmium, should not be utilized in direct burial service. Such metallic coatings are intended for the mitigation of atmospheric-type corrosion activity on the substrate steel.

6.6.2 Cathodic Protection

Mitigation of corrosion in onshore pipelines is primarily accomplished by the combination of cathodic protection and dielectric coating systems. The design of such cathodic protection systems is reasonably straightforward, because the corrosion engineer can often predict the distance at which protective current application from a remote anode bed will effectively protect the pipeline in both directions from a current drain point attached to the pipeline steel. The distance of effective full protection can be estimated if the pipeline diameter, steel type, wall thickness, soil characteristics, and general coating quality are known. This is so because a pipeline has an attenuation characteristic to current pickup from the electrolyte and longitudinal flow that is analogous to a leaky electric transmission line.

Cathodic protection of pipelines in soil is based on two general principles: (1) steel corrodes because portions of the material in the soil are anodic and others are cathodic, and (2) corrosion will not occur if all portions of the steel are cathodic. This is accomplished by impressing a direct electric current on the pipe and providing an anode that will corrode instead. This will not only reduce corrosion, it will stop it.

Pipelines are routinely protected by a coating supplemented with cathodic protection. An impressed current cathodic protection system for a pipeline would consist of a DC power source, which is often an AC powered rectifier and an anode or an array of anodes buried in the ground (the anode ground-bed). The DC power source would typically have a DC output of between 10 and 50 A and 50 V, but this depends on several factors, such as the size of the pipeline. The positive DC output terminal would be connected via cables to the anode array, while another cable would connect the negative terminal of the rectifier to the pipeline, preferably through junction boxes to allow measurements to be taken (Peabody, 2001). Anodes can be installed in a vertical hole and backfilled with conductive coke (a material that improves the performance and life of the anodes) or laid in a prepared trench, surrounded by conductive coke and backfilled. The choice of grounded type and size depends on the application, location, and soil resistivity (Peabody, 2001). The output of the DC source would then be adjusted to the optimum level after conducting various tests, including measurements of electrochemical potential.

Chemically, a measure of protection can be offered by driving a magnesium rod into the ground near the pipe and providing an electrical connection to the pipe. Since the magnesium has a standard potential of -2.38 V compared to -0.41 for iron, it can act as an anode of a voltaic cell with the steel pipe acting as the cathode. With damp soil serving as the electrolyte, a small current can flow in the wire connected to the pipe.

Offshore pipelines are usually protected from seawater corrosion by a coating, which is supplemented with cathodic protection to provide protection at coating defects or *holidays*. In the Gulf of Mexico, the pipeline coatings used until the early to mid-1970s were either asphaltic/aggregate, somastic-type coating (an extrudable mastic consisting of oxidized bitumen with mineral fill materials) or hot-applied coal tar enamels. Since then, the trend has been to use fusion-bonded epoxy powder coatings.

In the earlier days, the trend in cathodic protection was to rely on impressed current systems. In the 1960s and early 1970s, zinc bracelet anodes attached to the pipe were widely used. However, more efficient aluminum alloys have surpassed zinc as the preferred material for off-shore galvanic anodes. There are, however, still some operators using impressed current systems and some using zinc anodes.

Virtually all new pipelines installed in the Gulf of Mexico are equipped with aluminum bracelet anodes. There are two basic types: (1) square shouldered and (2) tapered. The square-shouldered anodes are typically used on pipe that has a concrete weight coating. When installed, the anodes are flush with, or slightly recessed inside, the outside diameter of the concrete. The tapered anodes are designed to be installed on pipelines with only a thin film corrosion coating.

When designing a cathodic protection system for a pipeline, the corrosion engineer has to consider the following variables, all of which will have an impact on the final anode alloy and size selection: (1) design life required—(minimum is 20 years), (2) pipe diameter length and to−from information, (3) geographic location, (4) type of coating, (5) water depth, (6) product temperature, (7) electrical isolation from platforms or other pipelines, (8) burial method, and (9) pipe-laying/installation method, which will have a direct impact on the amount of coating damage; there is also the risk of having anodes detached during the pipe-laying process.

Higher corrosion rates can be generally expected when the pipe coating has a combination of large damaged areas and adjacent pin-hole defects, and when the pipe is exposed to seawater rather than mud. There is also a particular risk of microbiologically influenced corrosion on buried lines with bitumen mastic-type coatings and depleted cathodic protection. If the cathodic protection systems have depleted, corrosion will begin at numerous sites all over the pipeline. Unless detected and retrofitted, the first leak could be the end of the pipeline, as the next several hundred won't be far behind.

Retrofitting the cathodic protection system with supplemental anodes would only make sense if the line in question is very old and the required additional life was significant. However, in addition to the cost to perform a pipeline cathodic protection, inspection will run up to several thousand dollars per mile; retrofitting pipeline cathodic protection systems offshore is not always a simple matter, especially when lines are deeply buried.

6.7 CORROSION CONTROL IN OFFSHORE STRUCTURES

Protective coatings are widely used in offshore structures because of the amount of carbon steel used in platform construction. A coating system usually consists of a primer, an intermediate or tie-coat, and a topcoat. An effective coating requires thorough surface preparation and skilled application in addition to the use of the best materials.

In terms of surface protection, the objective is to remove mill scale, to clean the surface of the steel, and to provide a suitable anchor pattern to ensure optimal bonding of the coating system. This is usually accomplished by surface sandblasting (Schremp, 1984).

When coating systems are employed, primers are used to protect the cleaned metal surface and are classed as either wash, zinc-rich, or inhibitive primers. Wash primers usually consist of a vinyl resin, solvent solution pigmented with zinc, or strontium chromate. Prior to application, the solution is mixed with phosphoric acid and alcohol. After application, the mixture produces a passive layer of iron phosphate on the metal surface. This film is easily damaged and should be top-coated quickly to prevent the formation of rust (Chapter 1).

Zinc-rich primers may be either organic or inorganic coatings with a high loading of zinc dust. These primers are excellent during construction because of their abrasion and impact resistance, but they are sensitive to acids and alkalis and should be protected with a chemically resistant topcoat when used on offshore platforms. Inhibitive primers may be thermoplastic resins (which cure by solvent loss) or thermosetting resins (which cure by addition of a catalyst or curing agent)—both systems include pigments, which are designed to retard corrosion in the presence of moisture either by ionization or by creating an alkaline environment at the metal surface. Both require topcoats to perform effectively. Intermediate and topcoats function as protective barriers for the primers by preventing the access of water, oxygen, and active chemicals. Both are usually of the same generic type and are made from thermoplastic or thermosetting resins.

Marine cathodic protection covers many areas, such as jetties, harbors, and (for the benefit of the present context) offshore structures. The variety of different types of structures leads to a variety of systems to provide protection. Typically, galvanic anodes are favored

(Roberge, 1999), but impressed current cathodic protection can also often be used.

Cathodic protection causes changes in the chemistry of seawater near the protected structure, and this causes the precipitation of a natural coating on the structure that reduces the need for a cathodic protection current.

The *deep draft caisson vessel* and SPAR-type structures have presented deep water operators with a cost-effective production facility for prospects with a moderate number of subsea wells and are being widely used in the Gulf of Mexico. These structures consist of the following main components, each of which has unique problems: (1) hull components, (2) tanks, (3), riser systems, and (4) miscellaneous areas.

6.7.1 Hull Components
The hull is typically a large cylindrical steel tube with a variety of internal compartments that provide buoyancy to support the topside drilling and production equipment with a central annulus (center well) through which all production is accomplished. Hulls are typically 90–120 ft in diameter with a center well diameter of 3050 ft. The overall length of the hull section can be up to or in excess of 600 ft. This construction presents two distinct seawater immersed corrosion exposures.

In the *outer hull*, in which normally basic grades of carbon steel are exposed to natural seawater, corrosion is prevented by using conventional marine epoxy coatings through the splash zone and topsides and conventional sacrificial anode cathodic protection to bare steel in the immersed region. Design criteria are the same as for conventional platforms. Monitoring of cathodic protection system performance is simply accomplished using surface deployed or remotely operated vehicle (ROV) interfaced monitoring equipment.

In the *center well*, the structure also uses carbon steel, which is exposed to quiescent seawater. The exclusion of normal sunlight reduces marine growth accumulation and densely packed riser systems make postinstallation access difficult to impossible, and can cause temperatures to riser to be a little above ambient seawater. The net result is a slightly lower current requirement for cathodic protection of the steel on the wall of the center well. Coatings are typically used only in the splash zone and emerged areas, and sacrificial anodes must be flush-mounted to avoid obstructions during riser installation—this fact must be taken into account when computing anode resistance.

There are also a series of guide frame structures within the center well; these guides restrict lateral movement of the riser systems during vessel relocation about its operational footprint and offset normal hydrodynamic forces. It is important to locate anodes on these guide frames, as they could be shielded from anodes on the inside wall. These anodes will provide a large proportion of current drained to the risers through fortuitous contact. The limited postinstallation access to this area makes it a good candidate for fixed reference electrodes to monitor the cathodic protection system performance.

Most recent designs utilize a *truss section* attached to the base of the main hull to support the bottom main ballast tank at the base of the SPAR structure. The truss is typically 300−325 ft (91−100 m) long. Construction is standard tubular steel members with periodic heave plates, which provide guides for the riser systems. The truss section is normally left uncoated and fitted with conventional sacrificial anodes.

The *main hull* will contain a number of internal compartments. Some of these are void tanks that contain only air and are never flooded. Other tanks are variable ballast tanks that will see raw seawater some of the time and will be largely dry for the majority of the life span. Being a part of the hull, these compartments are fabricated from basic grades of carbon steel. It is not uncommon for these tanks, particularly the variable ballast tanks, to be heavily baffled and reinforced, thus creating a large number of shielded compartments.

6.7.2 Tanks

Void tanks are usually vented to the atmosphere and thus cannot be fully deoxygenated. They are often susceptible to condensation on the inner walls. Access is possible through man-ways, and periodic visual and nondestructive examination is always a part of the required in-service inspection program. Access to effect coating repairs during operation is possible, but economically undesirable.

Coating is the main control against corrosion—possibly supplemented by vapor phase control as a backup method. Where allowed, coal tar epoxies fill the role because they are generally surface preparation tolerant.

Dehumidification can be a very effective method of void space corrosion control. The general adage, no water = no corrosion holds true and, should the dehumidification system not be 100%, vapor phase control as a secondary method of corrosion control is a reasonable strategy. Coupons for testing should be retrievable (preferably without tank entry) for monitoring corrosion rates.

For *variable ballast tanks* (*hard tanks*), coatings and sacrificial anode cathodic protection are required in most cases. It is important to locate anodes in all compartments and distribute them preferentially lower in the tank because of surface wetting. Particular attention should be paid to the coating quality in the upper areas of the tank likely to be in wet atmospheric service for much of the time; in these areas it is worthwhile considering the use of metallic-based primers or even thermally sprayed aluminum as a standalone coating or as an epoxy primer.

For *permanent ballast tanks*, construction is typically in the basic grades of carbon steel. The external surfaces of these tanks are simply left bare and cathodically protected. The inside surfaces present a fairly unique problem.

In order to provide maximum negative buoyancy at this location on the structure, it is necessary to fill the tank—not only with seawater but also with additional dense material. The ballast of choice is typically magnetite (Fe_3O_4) in a granular form. It is recommended that anodes be used above *and* below the magnetite, and the tank should be coated—particularly where the magnetite contacts the steel. Reference electrodes, as well as current density and anode current monitors, should be installed both above and below the magnetite surface.

6.7.3 Riser Systems

The riser systems (as opposed to the riser pipe on a refinery catalytic cracking unit) on this type of structures are referred to as *top-tensioned risers* (Bai and Bai, 2005). They are actually small structures within a structure, having their own buoyancy systems to support them. The only designed contact to the hull structure is through the topside flow-line connections that are above water. However, there is a very high probability of fortuitous contact through mechanical interference with the hull. The riser systems are free to move completely independently of the hull.

The *main riser sections* are of pipe-in-pipe type construction with the outer pipe acting as a conductor. At the base of the riser, there is a stress joint (there may also be a similar section near the point of exit at the base of the hull—referred to as the keel). The annulus between the riser and the outer pipe is usually flooded with seawater. Mechanical spacer centralizers are clamped to the inner riser pipe to control movement within the conductor pipe.

Air buoyancy cans are large tank-like structures built around the outer surface of the riser conductor near the top of the riser. Even though there are various designs, they have some common corrosion areas irrespective of specific design.

Outer can surfaces see the same environment as the center well areas of the hull, but buoyancy is critical and weight loading must be minimized. Anodes are not a good choice because of the additional weight as well as the possibility of mechanical interference with the risers' guide frames. The stroke length on these risers can be as much as 40 ft or more. While mechanical interference is absorbed on wear strips on the outside of the can, there is the good possibility of coating damage during installation of the risers.

Thermal sprayed aluminum provides a good solution in this area and, while serving primarily as a barrier coating the thermally sprayed aluminum, can also provide an adequate level of cathodic protection to small areas of exposed steel. However, the coating is conductive and will generally be at a potential that is 50 mV or more positive than the anodes in the hull center well, and will drain a small amount of cathodic protection from those anodes when contact between the structures exists.

Inner can surfaces are, during normal operations, mainly void, but they may have an open bottom that is exposed to seawater. Depending on the oxygen concentration, it may be necessary to coat the inside surfaces and provide a limited number of anodes at the base of the cans. Typically, the cans are filled with nitrogen in order to exclude the seawater during installation; periodic refilling is recommended to keep oxygen concentration to a minimum.

Export risers and flow lines from remote wells are either steel catenary design or may utilize flexible sections. Various methods have been used successfully utilizing either conventional fusion bond epoxy coatings or thermally sprayed aluminum. Anodes can be located on the riser catenary section, but some operators prefer to use anodes located at the touch down area and on the hull to provide protection from the ends.

6.7.4 Miscellaneous Areas

The *outer surfaces* of conductors are exposed to seawater (except inside the air cans) and see all resistivity layers in the seawater as they transit almost the entire water column. The use of conventional coatings with bracelet anodes has several drawbacks: (1) the possible shielding and coating damage under clamped stabilizer strakes and buoyancy modules, (2) the possible resistive buildup through mechanical joints, and (3) weight limitations. For these reasons, most systems use sealed thermally sprayed aluminum as the corrosion control.

The outside of the actual riser pipe should be treated like the outside of the conductor (and be coated with thermally sprayed aluminum). The inner wall of the conductor can be left bare if seawater in the annulus is suitably inhibited.

Stress joints are designed to take the major share of bending moment on the riser and are therefore made from high-strength materials with good stress characteristics and are located in the outer conductor pipe. Some titanium grades are particularly suitable from a mechanical standpoint and are therefore a common choice in many systems. The propensity of titanium to suffer hydriding under cathodic protection at certain levels provides a dilemma, one made more difficult by the lack of long-term field data and the very high consequence of a failure.

In deep water, it is common practice to predrill a number of *wells*, then temporarily cap them until the production structure can be located onsite. Some of these predrilled wells have nowhere to attach anodes for corrosion protection. In these cases, it is prudent to install a pod of anode material that can be electrically tied back to one or more of these wells using clamps.

Lateral *mooring systems* are required to enable the hull structure to be moved around its operational footprint. These multileg systems are subject to very high loading and a corrosion failure could be very costly. Cathodic protection is a suitable method for corrosion control, but the following precautions are required: (1) ensure that a flexible continuity jumper cable is provided between the fairlead structure and the hull and (2) ensure that current losses to the chain are calculated when sizing anodes.

6.8 CORROSION MANAGEMENT

The key to effective corrosion management is information, since it is on the basis of that information that ongoing adjustments to corrosion control are made. Information is valid data. Thus, to make effective corrosion management decisions on a day-to-day basis, the monitoring data must be valid. This is not simply a requirement for the probes to be operating correctly. It requires that they be placed in the most appropriate places, that is, at those points where the corrosion controlling activity might be expected to work, but where it might equally be expected to be least effective, for example, remote from the inhibitor injection point.

In many cases specially designed traps are introduced into a plant so that corrosion probes may be inserted. These often produce their own microenvironment, atypical of the plant itself, and with little hope of effective entry for an inhibitor. Data from a probe in such a location is unlikely to be relevant to corrosion management elsewhere in the system. Invalid data leads to ineffective corrosion management.

From time to time a corrosion management program should be reviewed at both the strategic and the tactical levels. In human affairs, things change. The management of a facility will always be alive to current market trends and it may be necessary to revise the management objectives from time to time. Since the corrosion management program was constructed to meet the objectives of an earlier plant management plan, it will be necessary to review the program and possibly to alter it. Likewise, the pace of technological change is rapid compared to the anticipated lifetime of most facilities. Thus, newer, more effective, cheaper means of achieving the same ends may emerge, and indeed, it may be possible to adopt them in place of existing tactics within the corrosion management program. Thus, the program is not a fixed blueprint, but a means to an end that must be reviewed and revised to meet the current management objective.

Corrosion cannot be ignored, for it will not go away. However, there is little merit in controlling corrosion simply because it occurs, and none in ignoring it completely. The consequences of corrosion must always be considered. If the consequences of corrosion can be tolerated, it is entirely proper to take no action to control it. If the consequences are unacceptable, steps must be taken to manage it throughout the facility's life at a level that is acceptable. To manage is not simply to control.

Good corrosion management aims to maintain, at a minimum life cycle cost, the levels of corrosion within predetermined acceptable limits. This requires that, where appropriate, corrosion control measures be introduced and their effectiveness be ensured by judicious, and not excessive, corrosion monitoring and inspection. Good corrosion management serves to support the general management plan for a facility. Since the latter changes as market conditions, for example, change, the corrosion management plan must be responsive to that change. The perceptions of the consequences and risk of a given corrosion failure may change as the management plan changes. Equally, some aspects of the corrosion management strategy may become irrelevant. Changes in the corrosion management plan must, inevitably, follow. For example, crude blending is the most common solution to high TAN crude processing. But blending can only be effective if proper care is taken to control crude oil and distillate acid numbers to proper threshold levels.

In order to manage corrosion, not only detection of corrosion is necessary but also an assessment of the extent of the corrosion (Nagi-Hanspal et al., 2013). Assessment of the extent of damage depends on inspection, or on an estimation of the accumulation of damage based on a model for damage accumulation, or both. Sound planning of inspections is critical so that the areas inspected are those where damage is expected to accumulate and the inspection techniques used are such as will provide reliable estimates of the extent of damage. If the extent of the damage is known or can be estimated, a change in strength can be ascribed to the component and its adequacy to perform safely can be calculated.

The general procedures for estimating fitness for service is outlined in the American Petroleum Institute (API) Recommended Practice 579—Fitness-for-Service guidelines, which provide assessment procedures for the various types of defects to be expected in pressurized equipment in a refinery. The steps are: (1) identification of flaws and damage mechanisms, (2) identification of the applicability of the assessment procedures applicable to the particular damage mechanism, (3) identification of the requirements for data for the assessment, (4) evaluation of the acceptance of the component in accordance with the appropriate assessment techniques and procedures, (5) remaining life evaluation, which may include the evaluation of appropriate inspection intervals to monitor the growth of damage or defects, (6) remediation if required, (7) in-service monitoring where a remaining life or inspection interval cannot be established, and (8) documentation, providing appropriate records of the evaluation made.

Creep damage can be assessed by various procedures and life. Life estimates can also be made based on the predicted life at the temperature and stress that are involved, by subtracting the calculated life used up, and by making an allowance for loss of thickness by oxidation or other damage. The growth of cracks in components operating at high temperatures that are detected can be estimated using established predictive methods (Webster and Ainsworth, 1994). Additionally, various examples of simplified methods to predict safe life in petrochemical plants containing cracks, for example in a reformer furnace, have been published (Furtado and Le May, 1996).

REFERENCES

Bai, Y., Bai, Q., 2005. Subsea Pipelines and Risers. Elsevier, Oxford, UK.

Bradford, S.A., 1993. Corrosion Control. Van Nostrand Reinhold, New York, NY.

Burlov, V.V., Altsybeeva, A.I., Kuzinova, T.M., 2013. A new approach to resolve problems in the corrosion protection of metals. Int. J. Corros. Scale Inhib. 2 (2), 92−101.

Bushman, J.B., 2002. Corrosion and Cathodic Protection Theory. Bushman & Associates Inc, Medina, OH.

Costerton, J.W., Geesey, G.G., Jones, P.A., 1987. Bacterial biofilms in relation to internal corrosion monitoring and biocide strategies. In: Proceedings of the Corrosion 87, NACE International. San Francisco, CA. March 9, Houston, TX.

Cross, C., 2013. High-acid crude processing enabled by unique use of computational fluid dynamics. Petroleum Technol. Q. Q4, 39−49.

Furtado, H.C., Le May, I., 1996. Damage evaluation and life assessment in high temperature plant: some case studies, in creep and fatigue. In: Penny, R.K. (Ed.), Ageing of Materials and Methods for the Assessment of Lifetimes of Engineering Plant. CRC Press-Balkema, Taylor & Francis Group, Leiden, The Netherlands.

Garverick, L., 1994. Corrosion in the Petrochemicals Industry. ASM International, Materials Park, OH.

Groysman, A., 1995. Corrosion monitoring in the oil refinery. Paper No. 07. In: Proceedings of the International Conference on Corrosion in Natural and Industrial Environments: Problems and Solutions. May 23–25, Italy.

Groysman, A., 1996. Corrosion monitoring in the oil refinery. In: Proceedings of the 13th International Corrosion Congress. November 25–29, Melbourne, Australia.

Groysman, A., 1997. Corrosion monitoring and control in refinery process unit. Paper No. 512. In: Proceedings of the Corrosion/97, New Orleans, LO.

Hau, J.X., Yépez, O.J., Torres, L.H., Vera, J.R., 2003. Measuring Naphthenic Acid Corrosion Potential with the Fe Powder Test. Rev, Metal Madrid Vol Extr, 116–123.

Hilton, N.P., Scattergood, G.L., 2010. Mitigater corrosion in your crude unit. Hydrocarbo Process. 92 (9), 75–79.

Jones, D.A., 1996. Principles and Prevention of Corrosion, second ed. Prentice Hall, Upper Saddle River, NJ.

Kane, R.D., Eden, D.C., Eden, D.A., 2005. Innovative solutions integrate corrosion monitoring with process control. Mater. Perform. February, 36–41.

Knag, M., 2005. Fundamental behavior of model corrosion inhibitors. J. Dispers. Sci. Technol. 27, 587–597.

Kuzub, V., and Novitskiy, V., 1984. Anodic protection and corrosion control of industrial equipment. In: Proceedings of the International Congress on Metallic Corrosion, Volume 1. National Research Council of Canada, Ottawa, Ontario, Canada, pp. 307–310.

Lack, J., Harrell, B., 2013. Reduce salt corrosion rates with stronger base amines. Hydrocarbon Process. 92 (9), 67–70.

Little, B., Wagner, P., Mansfeld, F., 1991. Microbiologically influenced corrosion of metals and alloys. Int. Mater. Rev. 36, 253–272.

Mottram, R.A., Hathaway, J.T., 1971. Some experience in the corrosion of a crude oil distillation unit operating with low sulfur North African Crudes. Paper No. 39. In: Proceedings of the Corrosion/71, NACE International, Houston, TX.

NACE., 2002. TM0497–2002, Measurement Techniques Related to Criteria for Cathodic Protection on Underground or Submerged Metallic Piping Systems, NACE International, Houston, TX.

Nagi-Hanspal, I., Subramaniyam, M., Shah, P., 2013. Corrosion control with high-acid crudes. Petroleum Technol. Q. Q4, 115–122.

Peabody, A.W., 2001. Control of Pipeline Corrosion, second ed. NACE International, Houston, TX.

Piehl, R.L., 1960. Correlation of corrosion in a crude distillation unit with chemistry of the crudes. Corrosion 16, 6.

Roberge, P.R., 1999. Handbook of Corrosion Engineering. McGraw-Hill, New York, NY.

Samuelson, G.J., 1954. Hydrogen–Chloride Evolution from Crude Oils as a Function of Salt Concentration, Proceedings. API, 34 (III), 50–54. American Petroleum Institute, Washington, DC.

Sastri, V., 1998. Corrosion Inhibitors: Principles and Applications. John Wiley & Sons, Hoboken, NJ.

Schremp, F.W., 1984. Corrosion prevention for offshore platforms. Soc. Petroleum Eng. J. 36 (4), 605–612.

Slavcheva, E., Shone, B., Turnbull, A., 1998. Factors controlling naphthenic acid corrosion. Paper No. 98579. In: Proceedings of the Corrosion/98, NACE International. Houston, TX.

Slavcheva, E., Shone, B., Turnbull, A., 1999. Review of naphthenic acid corrosion in oil refining. Br. Corrosion J. 34 (2), 125–131.

Srinivasan, V., Subramaniyam, M., Shah, P., 2013. Processing strategies for metallic and high-acid crudes. Petroleum Technol. Q. Q4, 51–57.

Stephen, M., 1986. Assessment of sulfide corrosion risks in offshore systems by biological monitoring. Soc. Petroleum Eng. J. 1 (5), 363–368.

Uhlig, H.H., Revie, R.W., 1985. Corrosion and Corrosion Control: An Introduction to Corrosion Science and Engineering, third ed. John Wiley & Sons, Hoboken, NJ.

Vetters, E., Clarida, D., 2013. Maintaining reliability when processing opportunity crudes. Petroleum Technol. Q. Q4, 59–67.

Videla, H.A., Herrera, L.K., 2005. Microbiologically influenced corrosion: looking to the future. Int. Microbiol. 8, 169–180.

Webster, G.A., Ainsworth, R.A., 1994. High Temperature Component Life Assessment. Chapman & Hall, London, UK.

GLOSSARY

Note: This chapter is available on the companion website: http://store. elsevier.com/product.jsp?isbn=9780128003466&_requestid=1050998.

Printed and bound by CPI Group (UK) Ltd, Croydon, CR0 4YY

03/10/2024

01040421-0012